A Decade of Middle School Mathematics Curriculum Implementation

Lessons Learned from the Show-Me Project

A volume in
Research in Mathematics Education
Barbara J. Dougherty, *Series Editor*

A Decade of Middle School Mathematics Curriculum Implementation

Lessons Learned from the Show-Me Project

Edited by

Margaret R. Meyer
University of Wisconsin–Madison

Cynthia W. Langrall
Illinois State University

With Associate Editors

Fran Arbaugh
University of Missouri–Columbia

David C. Webb
University of Colorado at Boulder

Murrel Brewer Hoover
WVSTEM Center @ Marshall University

IAP

INFORMATION AGE PUBLISHING, INC.
Charlotte, NC • www.infoagepub.com

Library of Congress Cataloging-in-Publication Data

A decade of middle school mathematics curriculum implementation lessons learned from the show-me project / edited by Margaret R. Meyer ... [et al.].
p. cm. – (Research in mathematics education)
Includes bibliographical references.
ISBN 978-1-60752-012-2 (pbk.) – ISBN 978-1-60752-013-9 (hardcover)
1. Mathematics–Study and teaching (Middle school)–Research. 2. Curriculum change. I. Meyer, Margaret Renee.
QA135.6.D43 2008
510.71'2–dc22

2008042386

Printed in the United States of America

This book is based on the work of the Show-Me Center supported by the National Science Foundation under grants ESI–9714999 and ESI–0137305. Any opinions, findings, conclusions, or recommendations herein are those of the author(s), and do not necessarily reflect the views of the National Science Foundation.

CONTENTS

CONTENTS

PART III

PROFESSIONAL DEVELOPMENT AND TEACHER EDUCATION

PART IV

WHAT HAVE WE LEARNED AND WHERE DO WE GO FROM HERE?

PREFACE

Hey, maybe we should write some of this down before we forget.
—Anonymous (after experiencing something significant)

PURPOSE AND AUDIENCE

The purpose of this book is to document the work of the Show-Me Project (1997–2007) and to highlight lessons learned about curriculum implementation. Although the Show-Me Project was charged with promoting the dissemination and implementation of four distinct comprehensive curriculum programs (*Connected Mathematics, Mathematics in Context, MathScape,* and Math*Thematics*), most of the lessons learned from this work are not curriculum specific. Rather, they cut across the four programs and share commonalities with standards-based curriculum reform at any level. We believe that documenting these lessons learned will be one of the legacies of the Show-Me Project.

But how does one portray lessons learned, if not through the voices of those directly involved in adopting and implementing these curriculum materials? Thus our story is told through the voices of those most intimately involved in these processes—curriculum developers, teachers, school district administrators, project directors, and researchers. We selected authors to tell very specific stories about the adoption and implementation of each of the four curricula. We asked district leaders to tell adoption and implementation stories based on their experiences with the four curricula. We enlisted teachers to write about changes in their instructional practice and beliefs about teaching mathematics that were influenced by

A Decade of Middle-School Mathematics Curriculum Implementation, pages ix–xv the curriculum materials they used. Developers contributed chapters that highlight aspects of their materials that should be the foundation of professional development involving their curricula. In addition to these curriculum-specific narratives, we sought authors to write about general issues pertaining to curriculum implementation, professional development for teachers and administrators, and the use of these materials in preservice teacher education.

We anticipate that the comprehensive nature of this work will attract readers from multiple audiences that include state and district mathematics supervisors, middle grades mathematics teachers and administrators involved in curriculum reform, as well as mathematics teacher educators. Those about to embark on the review of curriculum materials will appreciate reading about the processes employed by other districts. Readers with interests in a particular curriculum program will be able to trace the curriculum-specific chapters to gain insights into how the design of the curricula relate to professional development, adoption and implementation issues, and teachers' personal experience using the curriculum materials. Individuals who provide professional development at the middle grades level will find chapters that they can use for both general and focused discussions. Teachers at all stages of implementation will recognize their own experiences in reading and reflecting on the stories of teacher change. Mathematics educators will find ideas on how these curricula can be used in the preparation of preservice middle grades teachers.

ORGANIZATION OF THE BOOK

We have organized the book in four parts, beginning with an overview of the genesis and work of the Show-Me Project and concluding with our reflections on future directions. It is the chapters in between, however, that tell the stories from which we can draw lessons learned. These lessons have been distilled from the varied perspectives of those who have been involved in the work of the Show-Me Project and are presented in a culminating chapter. In organizing this volume we focused on both the general and the specific. Readers will find a number of chapters that address general issues related to all of the curricula. They will also find chapters specific to each of the four curricula supported by the Show-Me Project: an adoption story, an implementation story, curriculum specific professional development considerations, and a teacher change story. The chapters for *Connected Mathematics* are: 5, 8, 15, and 21; for *Mathematics in Context*: 6, 9, 16, and 22; for *MathScape*: 7, 10, 17, and 23; for MATH*Thematics*: 4, 11, 18 and 24.

Part I

Three chapters comprise Part I and provide the general reader with insights into the events that led to the establishment of the Show-Me Project and detail the specific activities over the ten years of the project. By way of introduction (Ch.1), Robert Reys situates the Show-Me Project within the historical context that led to its inception. This context includes the development of the NCTM *Standards* and the funding of standards-based curriculum development projects. In Chapter 2, Spud Bradley describes the events leading to NSF's decision to solicit proposals for implementation centers to support the dissemination and implementation of NSF-funded curricula. He discusses the contributions of the Show-Me Project from his perspective as the NSF program officer for the project. In Chapter 3, Barbara Reys documents the history of the Show-Me Project and describes its goals, activities, and accomplishments. Reys recounts the challenges faced by the Center and Satellites and addresses some of the politically motivated issues surrounding curriculum implementation. This chapter sets the stage for the presentation of lessons learned that follows in Parts II and III.

Part II

This part of the book is organized in two themes: *Issues of Curriculum Adoption* and *Contexts for Implementation.* The four chapters focused on adoption present different perspectives on the process of reviewing and adopting curriculum materials and highlight why this process is different for standards-based materials than for traditional curriculum materials. These chapters attempt to capture the variety of issues that districts might encounter. In Chapter 4, Cary Tuckey describes how MATH*Thematics* was adopted in a small school district where teachers were involved in a three-year professional development project aimed at changing teachers' instructional practice through the use of standards-based curriculum materials. Tuckey highlights two key components of the adoption process: the importance of piloting materials proposed for adoption and the value of incorporating a component of consensus building as part of the adoption process. In Chapter 5, Cheryl Rectanus describes the adoption of *Connected Mathematics* in a large urban district. She outlines an adoption model aimed at garnering consensus among all members of the adoption committee. Rectanus explains how the thorough work of the adoption committee enabled the district to weather negative comments reported in the press immediately following the adoption. In Chapter 6, Leslie Nielsen describes the process by which *Mathematics in Context* was adopted in a moderate sized district. She presents a detailed overview of the work of the program review committee and highlights the importance of keeping all teachers apprised of the adoption process. A noteworthy component of the adoption model described by Nielsen is the solicitation of feedback from students and their parents. And finally, in Chapter 7, Murrel Hoover reports on a statewide project designed to develop a cadre of teacher leaders, experienced in the use of *MathScape*, who would later be instrumental in directing and supporting the adoption and implementation of standards-based middle school curriculum materials in their districts. Hoover's chapter presents a unique approach to textbook adoption with its emphasis on systematic, widespread professional development prior to the adoption process.

The second section in Part II is comprised of six chapters focused on contexts for implementation. These chapters will be of special interest to district people responsible for guiding the implementation of standards-based curricula at any grade level. Chapters 8 through 11 present curriculum-specific implementation stories that detail the successes and challenges of implementing the four curriculum programs. In Chapter 8, Chip Sharp discusses the implementation of *Connected Mathematics.* He describes how the teachers formed a community that supported each other through the implementation. In Chapter 9, Carol Den Otter recounts the implementation of *Mathematics in Context* from her perspective as a district-level curriculum coordinator. Her chapter underscores the importance of supporting and sustaining implementation so that the effectiveness of the curriculum materials can be documented. Den Otter addresses issues such as how teachers handle being on the "front line" when the implementation is criticized by parents, the importance of teacher buy-in, and preparing teachers to assume leadership roles in the implementation process. In Chapter 10, Emily Fagan and Rhonda Weinstein present an implementation story about *MathScape.* Their narrative stresses the importance of teacher leaders in sustaining implementation after the initial year and the authors offer suggestions for addressing the challenges created by changes in staff in subsequent years of implementation. In Chapter 11, Linda Tetley details her work providing professional development in a district implementing MATH*Thematics.* She documents the specific concerns of teachers and shows how it is possible to recover when implementation gets off to a rocky start. Tetley illustrates the benefits of having professional development provided by someone experienced in teaching with the materials.

Chapters 12 and 13 are more general in nature. In Chapter 12, Jane Swafford and Cynthia Langrall present "hypothetical" stories of some of the particularly difficult challenges districts have faced when implementing these standards-based materials. The hypothetical cases they relate are based on a series of curriculum-specific case studies conducted by the authors and include specific issues identified by each of the curriculum developers. In Chapter 13, Jennifer Bay-Williams and Rebecca McGraw discuss ways of addressing critical implementation issues common to all standards-based curricula: language/literacy issues, and the notion of mathematics of *all* students. They provide recommendations based on research and grounded in the current related literature.

Part III

In this part of the book, we focus on *Teacher Education and Professional Development* as they relate to the content of these four curricula. In Chapter 14, Elizabeth Phillips, Glenda Lappan, Yvonne Grant, and Fran Arbaugh provide an overview of professional development needed for successful implementation through an examination of five contexts for professional growth: experiencing, planning, teaching, reflecting on, and professionalism. In Chapters 15 through 18, developers of the curricula discuss important elements of planning professional development that are specific to their curricula. In Chapter 15, Elizabeth Phillips and Glenda Lappan show how a problem-centered curriculum like *Connected Mathematics* built around significant mathematical ideas, teacher reflection, and the lesson structure of launch, explore, and summarize provides a framework for teachers to deepen their own understandings in preparation for teaching. In Chapter 16, David Webb emphasizes the role of context, models, and progressive formalization to outline professional development essential for teaching with *Mathematics in Context.* Fred Gross and Emily Fagan outline professional development considerations related to *MathScape* in Chapter 17. They discuss how a focus on context, communication, the design of lessons and units, and assessments prepares teachers to be successful in implementing the curriculum. In Chapter 18, Rick Billstein and Jim Williamson talk about professional development for MATH*Thematics* related to content, instructional approach, the use of technology, and assessment. In Chapter 19, Fran Arbaugh discusses how a dynamic professional development program needs to evolve to meet the changing needs of teachers involved in implementing these curricula. In Chapter 20, Margaret Meyer and Fran Arbaugh shift the focus to what administrators need to know about the curricula and professional development in order to support the adoption and implementation of standards-based curriculum materials.

In Chapters 21 through 24, teachers representing each of the four curricula write about how using the curricula in their classrooms resulted in changes in their teaching and understanding of mathematics. As a teacher of *Connected Mathematics,* Jane Porath talks in Chapter 21 about how teaching the curriculum deepened her own understanding about mathematical connections and how her expectations for student communication of mathematical understanding increased over time. In Barbara Dilieggio's chapter (22), she talks about the importance of collegial support when implementing *Mathematics in Context* (MiC) and how her own learning with MiC mirrored that of her students. In her chapter (23), Kelly Watts tells how teaching with *MathScape* enabled her to reach more students and to finally realize her goal of making a difference as a teacher. Fabienne Anderson tells of her experiences with teaching MATH*Thematics* in Chapter 24. Her story focuses on how the curriculum allowed her to grow as a teacher and the transformation she witnessed in her students. Chapter 25 is written from a mathematician's point of view. Continuing the theme of growth, Ira Papick tells how his experiences with these curricula impacted his views on teaching and learning mathematics, and how he came to collaborate to write a university-level curriculum for middle school preservice teachers.

Chapters 26 and 27 focus on preservice teacher education. In Chapter 26, Jeff Frykholm talks about how he uses the curricula in his methods course to support the examination of pedagogical and content knowledge. In Chapter 27, Ann Bledsoe provides a story about how these curricula can be used as a tool to help preservice teachers see how the theory proposed in their methods classes can come to life in middle school mathematics classrooms.

Part IV

We conclude the book with two chapters. In Chapter 28 Barbara Reys synthesizes key issues raised by the authors of the various chapters in parts II and III. Reys summarizes the lessons that have been learned through the work of the Show-Me Project and that were articulated by the authors. Finally in Chapter 29, we reflect on the challenges for the future of these middle grades curricula with regard to the lessons learned from the Show-Me Project. This chapter addresses some important issues that may not have been (or couldn't be) discussed in individual chapters.

ACKNOWLEDGMENTS

Many people made this book possible, but a few people deserve special recognition.

- The National Science Foundation, for the two grants that funded the ten years of the Show-Me Project.
- Our NSF program officer Spud Bradley, whose guidance and quiet encouragement let us know that this book project and all the work of the Show-Me Project was worth doing.

- The Show-Me director Barbara Reys, for 10 years of leadership and for her idea of a book to capture the lessons learned.
- The Satellite Directors Betty Phillips, Rick Billstein, Meg Meyer, and Fred Gross for helping us identify districts and teachers and for helping them write about their curriculum.
- Our reviewers Mary Lindquist, Eric Robinson, and Hannah Slovin, who gave the book a careful read and shared the wisdom of their experiences with reform curricula.
- The series editor, Barbara Dougherty, for recognizing the value of a book like this and for pitching the idea to the publisher.
- And most especially our Associate Editors Fran Arbaugh, David Webb, and Murrel Hoover for their time, efforts, insights, and general good humor in helping to conceptualize, write, and edit the manuscripts. Working with you has been a privilege and a pleasure.

Margaret R. Meyer
University of Wisconsin-Madison

Cynthia Langrall
Illinois State University

PART I

SETTING THE STAGE

The three chapters in this section provide background to the Show-Me Center Project. The first chapter by Bob Reys begins with the need for curricular reform in mathematics that resulted in the NCTM *Standards* (National Council of Teachers of Mathematics, 1989) and NSF's funding of the standards-based curricula and brings us to the beginning of the Show-Me Project. Chapter 2 by Spud Bradley of the National Science Foundation describes the events leading up to and resulting in NSF's decision to solicit implementation centers to support the dissemination and implementation of the NSF-funded curricula. Chapter 3 by Barbara Reys provides an historical overview of the Show-Me Center Project and describes the organizational structure of the Project in terms of the Center and the Satellites. She describes the goals of the project and how they were accomplished with specific discussion of Show Me activities.

CHAPTER 1

THE ROAD TO REFORM

Robert Reys
University of Missouri, Columbia, Missouri

Those who cannot learn from history are condemned to repeat it.
—George Santayana (1863–1962)

History provides valuable insights and perspectives to those wise enough to study the lessons offered and to learn from them. This truism applies to many fields including mathematics education. An important legacy from the "new mathematics" era of the 1950–1960s was the documentation of challenges and accomplishments of the School Mathematics Study Group (SMSG). In the book, *SMSG: The Making of a Curriculum*, Wooten (1965) provides an insider's look at a major mathematics curriculum project as it evolved. It is a fascinating story that documents the challenges of developing and implementing new curriculum models. Interestingly, some of those challenges mirror the same issues faced in the recent era of curriculum reform. This chapter provides the background and context for the establishment of the Show-Me Center. Two key actions formed the basis for the Center—first, the response of the National Council of Teachers of Mathematics (NCTM) to national calls for improvement to mathematics education and second, the National Science Foundation's (NSF) effort to create high quality curriculum materials and support teachers in improving their instructional practices.

A Decade of Middle-School Mathematics Curriculum Implementation, pages 3–8

NCTM STANDARDS LAUNCH A MAJOR REFORM EFFORT IN THE U.S.

During the 1980s, a succession of reports, such as *A Nation at Risk* (National Commission on Excellence in Education, 1983) and *Everybody Counts* (National Research Council, 1989), painted a bleak picture of the quality of education in American schools. More specifically, the reports raised concerns about the status of mathematics education and called for substantive changes in the teaching and learning of mathematics. In response to these concerns, the National Council of Teachers of Mathematics published the landmark document, *Curriculum and Evaluation Standards for School Mathematics* (NCTM, 1989). The *Standards* represented a precedent-setting document providing a fresh vision for mathematics curriculum. Never before had a professional organization tackled the problem of offering a comprehensive vision for school mathematics. The *Standards* and later the *Professional Standards for Teaching Mathematics* (NCTM, 1991) and *Standards for Mathematics Assessment* (NCTM, 1995), included recommendations that called for significant changes in mathematics—both in content and pedagogy (Reys, Robinson, Sconiers & Mark, 1999). The recommendations included a call for *all* students to learn important mathematics. They also emphasized the value of curriculum and instruction that engages, challenges, and prepares students for continued study and growth in mathematical skills and understandings, causes them to develop mathematical habits of mind, and develops an understanding and appreciation of the important role of mathematics in scientific applications in their lives and as an interesting discipline of study.

The intent of the *Standards* was for students to become mathematically literate, including being able to explore, to conjecture, to reason logically and to use a variety of mathematical methods to solve problems. From these documents and the professional conversations that ensued, the phrase "standards-based reform" became a rallying cry. While these documents did not endorse or recommend a particular set of curriculum materials or teaching model, they provided guidance and argued strongly for the need to improve school mathematics programs.

Many teachers embraced the vision of school mathematics offered by NCTM and actively worked in their classrooms to realize this vision. However, they often did so by substantially modifying or adapting their school mathematics textbook. In fact, in many reform classrooms, the district-adopted textbook was set aside as it did not fit the new approach to mathematics instruction (Acquarelli & Mumme, 1996). Clearly, there was a need for new types of curriculum materials to support NCTM's vision. NCTM responded by producing an "addenda" series to the *Standards* that provided activities on a variety of topics and at different grade levels. The popularity of the Addenda series provided evidence that teachers were seeking new kinds of curriculum materials. However, few mainline publishers offered such materials as basal texts. Thus, many teachers supplemented their district-adopted textbooks with "replacement units" drawn from innovative resources from publishers such as Dale Seymour Publications, Creative Publications, and Marilyn Burns and Associates. However, this represented a band-aid approach to curriculum improvement. That is, rather than developing new curriculum materials—based on the *Standards* and current research—to address the needs of students, supplemental resources were used in conjunction with existing, traditional textbook materials. The result was a patchwork of mathematics resources that often lacked a coherent structure.

NSF FUNDS 13 COMPREHENSIVE MATHEMATICS CURRICULUM DEVELOPMENT PROJECTS

Given the vacuum of existing mathematics curriculum materials (textbooks) that were consistent with "standards-based reform" and the reluctance of commercial publishers to invest large sums of money developing new mathematics programs whose ability to sell and gain significant market share (and thereby turn a profit) were unknown, the National Science Foundation "primed the pump" by investing about $70 million over a 6-year period in the development of mathematics curricula that reflected standards-based principles. Three NSF solicitations were announced during 1989–91 calling "for the creation of improved materials and model programs for mathematics instruction" (see Chapter 2 by Bradley).

A requirement of the solicitations included collaborative partnerships with commercial publishers so that the materials developed could be widely disseminated. Collaboration with commercial publishers was a precedent-setting initiative by NSF. It reflected one of the lessons learned from the 'new mathematics' era of the 1960s, namely that if mathematics curricula are to be widely used in schools across the United States, then it is essential that commercial publishers, including their marketing and sales force be utilized to support widespread impact. As a result of NSF solicitations, 13 comprehensive mathematics curriculum development projects were funded (see Table 1.1 for a list of projects, sponsoring institutions, and initial publishers).

These comprehensive, multi-grade, standards-based mathematics curricula funded by NSF came on the market in the mid- to late-1990s through

TABLE 1.1 Comprehensive Mathematics Curriculum Development Projects Funded by NSF from 1990–1996

Curriculum Name	Grades	Curriculum Development Sponsor	Initial Commercial Publisher
Investigations in Number, Data, and Space	K–5	TERC	Dale Seymour Publications
Math Trailblazers	K–5	University of Illinois-Chicago	Kendall/Hunt Publishing Company
Everyday Mathematics	K–5	University of Chicago	Everyday Learning Corporation
MATH *Thematics*	6–8	University of Montana	McDougal Littell
MathScape	6–8	Educational Development Center	Creative Publications
Mathematics in Context	5–8	University of Wisconsin	Encyclopaedia Britannica
Connected Mathematics Project	6–8	Michigan State University	Dale Seymour Publications
Pathways to Algebra and Geometry	6–8	Institute for Research on Learning	Voyager Expanded Learning
Contemporary Mathematics in Context	9–12	Western Michigan University	Everyday Learning Corporation
MATH Connections	9–12	MathConx	*It's About Time*, Inc.
SIMMS Integrated Mathematics	9–12	Montana Council of Teachers of Mathematics	Simon & Schuster Custom Publishing
Interactive Mathematics Program	9–12	Sonoma State University	Key Curriculum Press
Mathematics Modeling Our World	9–12	COMAP	South-Western Educational Publishing

their commercial publishers. Although each curriculum is different, all provide an interpretation of the vision outlined by the NCTM *Standards* (1989). That is, they:

- Introduce and connect significant mathematical concepts, consistent with the curriculum of other high achieving Trends in International Mathematics and Science Study (TIMSS) countries;
- Emphasize mathematical understanding, problem solving and reasoning;
- Use applications and learning contexts that interest, challenge, and motivate student investigation;
- Provide all students with opportunities to learn important mathematical ideas and skills;
- Provide opportunities for significant interactions among students and teachers utilizing a variety of communication methods (writing, explaining, presenting, defending);
- Incorporate use of technology that enhances understanding;
- Provide a rich source of materials and a vehicle for teachers at various professional development entry levels to utilize and learn from;
- Utilize various forms of alternative assessment to better gauge student understanding and facilitate instructional practice;
- Build a strong base for various high school mathematics options (Reys, Robinson, Sconiers & Mark, 1999).

Funding the development of numerous comprehensive curriculum projects represented a substantial and long-term commitment by NSF. The development efforts included time to conceptualize the curricula, develop materials, pilot them in schools, and revise to reflect experience gained from classroom use. This model of develop—pilot—revise is virtually unheard of in the current commercial textbook publishing industry. Instead, commercially developed textbooks are constructed rapidly to meet short deadlines established by state-adoption regulations and therefore preclude extensive piloting and revision prior to publication (Tyson-Bernstein, 1988; Finn & Ravitch, 2004; Reys & Reys, 2006).

Few envisioned that the standards-based mathematics curricula would, in and of themselves, revolutionize school mathematics programs. However, many felt that the existence of these important curriculum tools would support teachers and administrators in offering a richer, more challenging curriculum to students. Early findings from the pilot materials suggested that they offered the potential to increase student understanding and problem solving while maintaining skill levels (Carroll, 1997; Zawojewski, Robinson & Hoover, 1999; Huntley, Rasmussen, Villarubi, Sangtong & Fey, 2000; Senk & Thompson, 2003). Early use also identified challenges with implementing new curriculum materials, including a need for initial and sustained professional development for teachers (Bay, Reys & Reys, 1999).

NSF FUNDS THE SHOW-ME CENTER

The new NSF-funded curriculum materials prompted considerable interest among teacher leaders and researchers. As expected, adoption of the materials was initially slow as few school districts were ready to commit to such widespread changes. However, through professional development initiatives including some initiated by NSF (e.g., state and local systemic change projects), increasing numbers of teachers were learning about the new curriculum materials and trying them in their classrooms.

In 1995 NSF issued a call for Centers to support implementation of standards-based curriculum materials. By 1997 NSF had funded four mathematics curriculum centers, one for each grade-level band (elementary, middle, secondary) and a K–12 Center, to promote the dissemination and implementation of these materials. And thus begins the story of the Show-Me Center, which was established to support the middle grades mathematics curricula and through its work nationwide, championed curriculum reform efforts for more than 10 years.

CHAPTER 2

THE GENESIS OF THE NSF CURRICULUM IMPLEMENTATION CENTERS

John S. (Spud) Bradley[1]
National Science Foundation, Arlington, Virgina

NSF invests in the best ideas generated by scientists, engineers, and educators working at the frontiers of knowledge, and across all fields of research and education.

—NSF Strategic Plan,
(http://www.nsf.gov/pubs/2006/nsf0648/nsf0648.jsp)

In the early 1990s, the National Science Foundation (NSF) funded 13 curriculum projects to develop mathematics instructional materials to reflect the recommendations of the *Curriculum and Evaluation Standards for School Mathematics* and *Standards for Teaching Mathematics* (National Council of Teachers of Mathematics (NCTM), 1989, 1991). In the mid 90s as the curricula began to be commercially available, NSF funded four projects to support the dissemination and implementation of these curricula. The Show-Me Center, housed at the University of Missouri, focused on dissemination and implementation issues associated with the five middle school curricula. COMPASS, housed at Ithaca College, dealt with the five high school curricula and the ARC Center, housed at COMAP (Consortium for Mathematics and its Applications, in Bedford, MA), focused on the three elementary grades curricula. The fourth center was at the Education Development

A Decade of Middle-School Mathematics Curriculum Implementation, pages 9–15

Center and was concerned with all 13 NSF-funded curricula. This chapter details the NSF perspective that lead to the funding of these centers.

LOOKING FORWARD TO IMPLEMENTATION

The first systematic discussion of curriculum implementation involving NSF staff and curriculum developers occurred at Gateways III in Palo Alto, California in October 1994. The Gateways meetings were held annually for the project teams that had received awards from NSF to develop curricula. These meetings were designed to foster exchanges of information and cooperation among the developers. The goals of the first two Gateways conferences had been to reduce the natural feeling of competition among the developers and to recognize the common challenges they were encountering and to explore ways of resolving them. The latter goal was an especially important consideration given the radical departure of these materials from the traditional mathematics materials.

During this meeting Margaret (Midge) Cozzens, who was then Director of NSF's Division of Elementary, Secondary, and Informal Education (ESIE), addressed development teams and urged them to begin thinking about what would be needed for schools to implement the materials they were developing. At this point in time, this message was not something the developers were ready to hear. They were still immersed in development, planning for pilot and field testing, providing professional development for pilot and field-test teachers, and trying to respond to NSF's admonitions about not seeing themselves as competitors. Although developers knew that implementation would be another critical issue, it was hard to plan for something that far in the future. Cozzens urged the developers to begin making plans for dealing with these issues. She announced that NSF would entertain proposals to address issues of dissemination and implementation. Since this was an entirely new area of focus for NSF and no models for the work existed, Cozzens provided few guidelines for the work. The one big caveat of the Request for Proposals (RFP) was that NSF would not fund curriculum development teams to implement their own materials. The reason for this caveat was twofold. First, the curriculum development grants were sizable and NSF was under pressure to spread out the implementation funding to other people. Secondly, it was thought that involving different people to focus on implementation would provide new perspectives on the finished curricula.

The first proposals were submitted in early 1995. Most of these proposals were from small consortia of the curriculum development teams. This was not acceptable to NSF and all the proposals, except one, in this first round were declined. The proposal that was funded was a project at COMAP (Sol

Garfunkel, Principal Investigator) to develop video and associated print materials that would provide a picture of what classrooms using these materials would look and feel like. Another project was funded that should be viewed as part of the implementation effort. This was a Small Grant for Exploratory Research made to Ithaca College (Eric Robinson. Principal Investigator) to analyze the emerging curricula and clearly identify the mathematics found in them. The rationale for this work was that one could not simply read the table of contents or flip through the pages of the materials to easily determine the mathematical content. Whereas one could simply look at the table of contents of a traditional Algebra I text and find headings such as Rational Expressions, or Exponents and Radicals, in the NSF-funded Integrated Mathematics Program (Fendel, Resek, Alper, & Fraser, 1997–1999) one would find headings such as "Pig" and "The California Experience." Robinson proposed to explicate the mathematics content in all 13 curricula.

STRUCTURING THE REQUEST FOR PROPOSALS

During this period, from late 1994 into early 1997, the nature of the needed implementation effort was not obvious. Both NSF staff and the development teams recognized that the materials being developed were dramatically different from those to which teachers were accustomed and that implementation needed special attention, but there was no agreement about how to meet this need. No one knew how well the materials would be received. While many teacher leaders, particularly those who were active in NCTM, knew of the new curriculum materials, it seemed highly likely that the vast majority of teachers were unaware of the availability of the new curricula.

Everyone agreed that efforts were needed to make a wider collection of teachers aware of the materials, but recognized that awareness was only the beginning. A different kind of pedagogy and classroom management would be required for the effective use of these materials. The mathematics was often embedded in contexts intended to engage student interest and the actual mathematics was not always evident from a cursory reading. Because of these features, more professional development would be required than was common with the adoption of a traditional text. However, neither the publishers who had signed up to publish the materials nor the school districts that were early adopters were aware of the extent of this need. Moreover, there were not large numbers of professionals who understood the materials and had the skills to provide the needed professional development. It was also recognized that there would be a need to educate parents about the changes in their children's math experiences. Nevertheless, there were still uncertainties about the balance between the need to develop awareness of the curricula and the more difficult work of helping teachers and administrators understand the professional development needs.

Conversations about the dissemination and implementation issues continued at NSF and there were informal exchanges with developers. NSF had invested about $70 million in the development of the 13 curricula and program staff believed that it was important that NSF continue to support the materials to ensure that they be tried out in a reasonable number of classrooms in order to determine their effectiveness.

A model for implementation centers emerged at the April 1996 NCTM meeting in San Diego. After a number of informal discussions, I met with several members of development teams, including Sol Garfunkel, Chris Hirsch, and Dan Fendel (all of whom were developing high school materials). At that time I presented the NSF view that the implementation effort had to be independent of the developers. Dan Fendel argued that any successful implementation effort would require knowledge of the design principles the developers employed and familiarity with the materials being implemented. He also stressed that only the members of the development teams had this knowledge. The other developers strongly agreed. After further discussion someone at the table suggested that a workable compromise might be a project at a central site led by someone not associated with the developers, but with "satellites" consisting of development teams. Since there was still at least a perceived feeling of competition among those teams developing materials at the same grade levels, everyone agreed that the central site would have to be managed by someone that all the participating developers perceived as being both impartial and knowledgeable about the nature of the materials.

There was immediate agreement that because of his analysis work with the curricula, Eric Robinson would be the ideal person to lead the high school implementation effort based on this center/satellite model. After additional discussion a final model emerged for an implementation project consisting of a central site at Ithaca College with satellite sites, through subcontracts, at each of the five high school curriculum development projects.

FUNDING THE IMPLEMENTATION PROJECTS

The official NSF interest in implementation projects was reflected in the ESIE program solicitation, NSF 95-150, in a short paragraph under *Special Emphases.* The RFP elicited a second round of proposals for implementation projects. The model that emerged in San Diego, as well as other proposed models, was responsive to the solicitation's special emphasis on implementation. This round of proposals was reviewed in the summer of 1996 and two implementation projects were funded. Ithaca College (Eric

TABLE 2.1 Summary of NSF-Funded Mathematics Centers

Name	Initial Funding	Sponsoring Institution	Director	Website
COMPASS	1997	Ithaca College	Eric Robinson	www.ithaca.edu/compass/
Show-Me	1997	U. of Missouri	Barbara Reys	showmecenter.missouri.edu
ARC	1998	COMAP	Sheila Sconiers	www.comap.com/elementary/projects/arc/
K–12 MCC	1997	EDC	June Mark	www2.edc.org/mcc/

Robinson, Director) was funded for a project focused on high school materials with each of the high school materials development projects included as a satellite through subcontracts from the center. The K–12 center funded at the Education Development Center (June Mark, Director) was funded to focus on implementation issues related to all 13 sets of materials being developed with NSF support. A middle grades center and an elementary center built on the same model were also proposed, but were not funded in this round.

The following year, Instructional Materials Development (IMD) program solicitation (NSF 97-20) contained a separate category for implementation projects. This call resulted in proposals for projects in both mathematics and science. Among the mathematics proposals was one from the University of Missouri for the Show-Me Center. The proposal presented an excellent plan based upon the center/satellite model that connected in appropriate ways with the developers of the middle grades curricula. The PI and Co-PIs were well qualified and had considerable experience with middle school mathematics materials. The reviewers recognized the strengths of the proposal and the Show-Me Center was funded for five years. The ARC Center at COMAP, built on the same model of a central site with developer satellites and with well qualified and experienced leadership, was also funded. Table 2.1 summarizes the four national Implementation and Dissemination Centers funded by NSF for mathematics.

THE EVOLVING WORK OF THE CENTERS

Even with four funded projects neither the NSF staff nor the implementation project staffs had a clear understanding of the balance needed between promoting awareness of the instructional materials and working with schools and districts to help them with selection, adoption, and implementation issues. Each of the four implementation projects hosted national "curriculum showcases" in the early months to familiarize key school personnel with the materials. During the first year or so, experience showed that although awareness was still important, it was becoming less important for the centers as more districts began to ask for assistance with implementation issues.

Annual reports, site visits, collaborations with large-scale systemic initiatives, and reports from the field all indicated that the implementation centers were effective in meeting their goals of supporting dissemination and implementation of the curricula. In order to further evaluate progress and plan for future work, NSF funded a conference in June 2001 involving the implementation centers and selected school personnel. The critical role of these projects in assisting school districts in a long-term process of helping teachers use the new curricula and in building local capacity to support longer-term efforts was confirmed at this meeting. Based on the available evidence, ESIE program staff concluded that the implementation projects were indeed effective and announced that proposals for renewal of the implementation centers would be accepted. All four of the mathematics centers applied and received funding for support beyond the original five-year period.

PROJECT RENEWALS LIMITED BY CRITICISM

NSF's ten-year involvement in funding the development and implementation of mathematics curricula was not universally lauded. In fact, criticism of the NSF-supported mathematics instructional materials had become intense from some quarters. In March 2002, the American Enterprise Institute hosted a forum on mathematics education that featured many of the leading critics of these instructional materials and of NSF's funding of them. Immediately after this event, the Director of NSF began to receive numerous and persistent email messages criticizing the NSF-funded mathematics materials and the implementation projects. During this same period there were leadership changes in ESIE and at the NSF directorate level as well as substantial budget reductions in the Instructional Materials Development and the Teacher Enhancement programs; the programs that shared the funding of the implementation projects. The persistent criticism had elevated the profile of the implementation projects. This and the budget cuts led to a decision to curtail the funding of the implementation project renewal proposals to only two years. The renewal of the Show-Me Center was an exception. Its renewal period was for three years (five years had been proposed) because of the central role of the middle school experience, the overall quality of the work done by the center, and the fact that the center leadership had made some very difficult decisions regarding continued support of one of the satellite centers.

FINAL THOUGHTS

As this is written in early 2006, I reflect on the remarkable impact made by the materials development effort of the early 1990s. Compared to the generally uniform nature of the texts traditionally available to schools, these materials contained new mathematical content in new and innovative formats. At the same time, they required dramatically different pedagogy and assessment regimes, placing greater cognitive demands on teachers and students, and required different methods of classroom management. Few would have predicted when this effort began that the materials would be so widely used (for example, *Connected Mathematics* [Lappan, Fey, Fitzgerald, Friel, & Phillips, 1998, 2002, 2004, 2006] has between 20% and 25% of the middle school market) or that most of the (few) large publishers would actively seek to have one of these curricula in their portfolio. There is no single reason for this success, but there is no doubt that the implementation projects have played a major role in this outcome. The Show-Me Center played an especially important role in the success of the middle grades curricula.

NOTE

1. Spud Bradley served as the NSF program officer for the Show-Me Project during the 10 years of the project. The project staff gratefully acknowledges his support and encouragement.

CHAPTER 3

THE SHOW-ME PROJECT

A Decade of Support for Standards-Based Curriculum Implementation

Barbara Reys
University of Missouri, Columbia, Missouri

This chapter provides a brief history of the Show-Me Project, a National Science Foundation (NSF) sponsored Implementation and Dissemination Center funded from 1997–2007, with partners that included five NSF-funded middle-grades mathematics curriculum development projects. The intent of this chapter, indeed this entire book on the Show-Me Project, is to capture some of its many dimensions, and thereby, contribute to the legacy of the major mathematics curricular reform of the 1990s. Peter Dow (1991) said, "History is the past seen through the eyes of others." In the case of this chapter, the history of the Show-Me Project is reported through my eyes as its Director and as such this history is my rendition of how the Show-Me Project came to be and its contributions to the field.

The Show-Me Project (SMP)[1] began with official notice of funding from the NSF in October 1997. However, the impetus that led to the establishment of the SMP began nearly a decade earlier. These seeds are summarized in Chapters 1 and 2 of this volume and include:

A Decade of Middle-School Mathematics Curriculum Implementation, pages 17–31

- The National Council of Teachers of Mathematics (NCTM) release of major curriculum recommendations in *Curriculum and Evaluation Standards for School Mathematics.* (1989)
- The National Science Foundation's initiative to support major comprehensive mathematics curriculum development projects. (1990–92)
- Distribution by commercial publishers of four comprehensive middle school mathematics textbook series. (1995–97)

The principal investigators of the NSF middle school mathematics curriculum projects began disseminating information about their work in 1993 during professional meetings. At about the same time, a state Eisenhower grant was awarded to the University of Missouri to introduce local middle school mathematics teachers to new models of curriculum materials. These teachers had expressed dissatisfaction with the mathematics textbooks they were currently using. They confided that their mathematics textbooks had too much repetition, and were generally uninteresting for both teachers and students. They reported too many middle school students were not intellectually challenged and consequently being turned off to mathematics. Rightly or wrongly, these teachers placed much of the blame for these problems on their mathematics textbooks. The Eisenhower Project, *Reshaping Mathematics in the Middle Grades: A Model to Improve Mathematics Curriculum and Teaching,* supported summer workshops in 1993 and 1994 designed to introduce the new middle school curriculum materials to local teachers. Invitations were extended to directors of four curriculum development teams—*Connected Mathematics Project* (CMP) (Lappan, Fey, Fitzgerald, Friel, & Phillips 1998, 2002, 2004, 2006), MATH*Thematics* (MT) (Billstein & Williamson, 1999a), *Mathematics in Context* (MiC) (National Center for Research in Mathematical Sciences Education & Freudenthal Institute, 1997–1998; Wisconsin Center for Education Research & Freudenthal Institute 2006), and *MathScape* (MS) (Education Development Center, Inc., 1998, 2005)—to introduce their materials to project teachers. This resulted in Betty Phillips (CMP), Rick Billstein (MT), Meg Meyer (MiC) and Dan Brutlag (MS) leading sessions over a period of two weeks. These representatives provided an orientation to the project goals and engaged participating teachers in some of the lessons being developed. All 25 of the teachers that participated in the two-week session became excited about these materials and the potential they had for their own classroom use. On their own, the teachers initiated plans to use many of these lessons as replacement materials during the next school year. Teachers from one school (Jefferson City Middle School) volunteered to serve as a pilot site for MATH*Thematics* and thus began a commitment to this curriculum that continues today (Chapter 11 by Linda Tetley in this volume tells that story. Also see Tetley, 1998).

The Eisenhower Project provided convincing evidence that curriculum investigation was a powerful model for engaging middle school mathematics teachers in professional development that would translate to classroom practice. Consequently, in response to a 1994 NSF Teacher Enhancement solicitation, a proposal was submitted to extend the model statewide. The proposal resulted in the Missouri Middle-grades Mathematics (M^3) Project.

The M^3 leadership team included mathematics educators and mathematicians (John Beem and Ira Papick, Professors in the Department of Mathematics at the University of Missouri). The project involved working with teacher/administrator teams from 23 different school districts in Missouri to help them become informed about the NSF-funded middle school mathematics curricula, use selected units in their schools, and discuss and make decisions about their appropriateness for their schools (Reys, Reys, Barnes, Beem, & Papick, 1997).

The Eisenhower and NSF Teacher Enhancement projects provided over 4 years of experience working with middle school mathematics teachers and witnessing the renewals that were happening as teachers became excited about using standards-based mathematics curriculum materials. School administrators were also buying into this model and spearheading work in their schools to provide a structure to support curriculum reform. Project leaders became convinced of the necessity of supportive structures and curriculum-based professional development in order to maximize the potential of standards-based curriculum.

NSF ESTABLISHES IMPLEMENTATION AND DISSEMINATION CENTERS

As the development of the mathematics curriculum projects led to published materials in the mid 1990s, NSF initiated a solicitation to establish Implementation and Dissemination Centers (see Spud Bradley's summary of this initiative in Chapter 2 of this volume). The intent of the centers was to provide support for school districts initiating major curriculum reform to support improvement of student learning.

In the fall of 1996, Glenda Lappan (on behalf of the five NSF middle school curriculum development teams) contacted me and encouraged me to lead an effort to submit a proposal for a Dissemination and Implementation Center for middle school mathematics. A proposal, *The Show-Me Project: A National Center to Support Dissemination and Implementation of Standards-Based Middle Grades Mathematics Curricula,* was submitted to the NSF and was officially approved with a start date of September 1997. Thus after years of curriculum development effort and state-level work to use standards-based mathematics curriculum investigation as a vehicle for professional development, the national Show-Me Project was born.

THE SHOW-ME PROJECT

The charge of the Show-Me Project was to disseminate information about the 5 NSF-funded comprehensive middle grades mathematics curriculum projects and to support teachers and schools in their efforts to implement these materials. It was originally funded for 5 years (1997–2002) and then renewed for another 3 years (2002–05). Due to no-cost extensions, it continued to operate for a total of 10 years (1997–2007). The originally stated broad goals of the Show-Me Project included:

1. Disseminate information supporting awareness, examination, and implementation of comprehensive standards-based middle school mathematics curricula.
2. Design, develop, and support professional development opportunities for middle grades teachers, administrators, and other key personnel committed to improving middle school mathematics programs through standards-based curricular and instructional innovation.
3. Provide communication networks for middle school teachers, administrators, parents, and guidance counselors as they consider and/or implement new curricular and instructional techniques designed to advance student understanding and achievement in mathematics.
4. Provide information on and opportunities for professional development to mathematics educators/mathematicians in higher education who are involved in the preparation of future generations of middle grades mathematics teachers.
5. Collect, organize, and disseminate information regarding standards-based mathematics reform at the middle grades level.

As the Show-Me Project evolved, it focused on three main themes: curriculum leadership development, teacher professional development through curriculum investigation and enactment, and dissemination of information on curriculum development, implementation and evaluation.

Show-Me Project Structure

The Show-Me Project included a Center at the University of Missouri which oversaw all aspects of the work and five Curriculum Satellite Centers, each the home of an NSF-funded middle school curriculum development project. In addition, an external team at Illinois State University provided ongoing evaluation of project activities. The fiscal home of the project was the University of Missouri. The five Curriculum Satellite Centers and the evaluation team were supported through subcontracts to their respective institutions.

Figure 3.1 provides a schematic of the organizational structure of the project and Table 3.1 summarizes a list of project staff. At each Center and Satellite site, local staff included site Directors, support staff and/or doctoral students. The Show-Me National Advisory Board included prominent mathematics educators, mathematicians and school leaders. The Board met annually to review the work of the project and provide advice. In 2002, twelve Show-Me Regional Associates and four Show-Me Master Teachers were identified to expand the outreach of the project. The expanded Show-Me leadership team met annually to plan and advance the work of the project.

Show-Me Project: Major Activities/Initiatives

As noted in Figure 3.1, in the first round of funding (1997–2002), the Show-Me Project consisted of representatives from five middle school curriculum development teams (Satellite Centers) and the staff (faculty and doctoral students) from the University of Missouri (Show-Me Center). Members of each team met frequently (3–4 times per year) to plan the

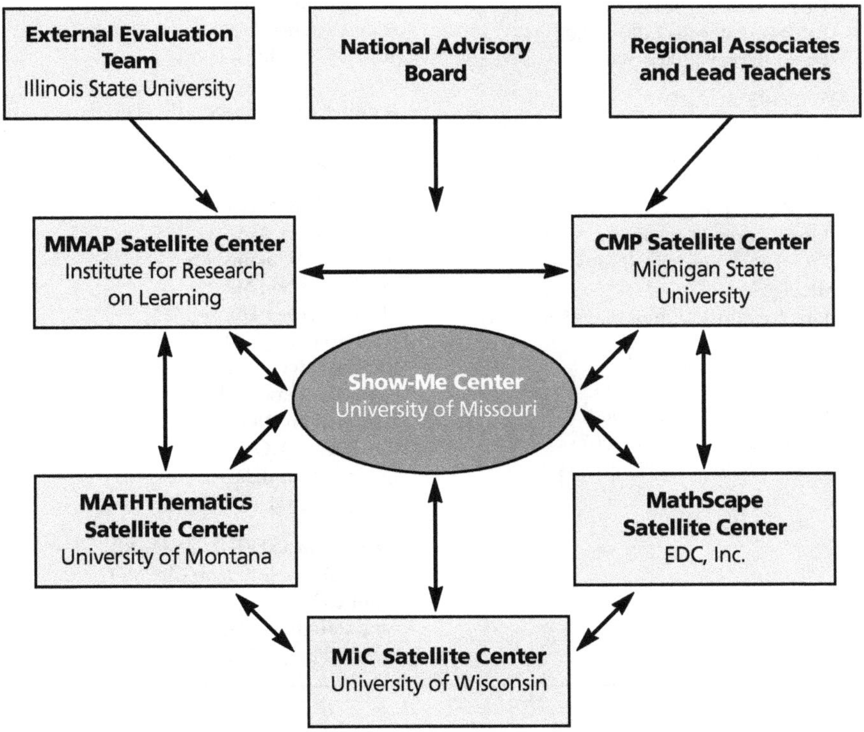

Figure 3.1 Organizational structure of the Show-Me Project.

TABLE 3.1 Show-Me Project Personnel (1997–2007)

SHOW-ME CENTER STAFF
University of Missouri, Columbia
Barbara Reys, Director (1997–2007)
Fran Arbaugh, Co-Director (2002–2007)
David Barnes, Co-Director (1997–2001)
Joy Whitenack, Co-Director (1997–1999)
Ira Papick, Co-Director (1997–2007)
Robert Reys, Co-Director (1997–2007)
James Tarr, Co-Director (2002–2003)

SHOW-ME POST-DOCTORAL FELLOW
Oscar Chavez (2003–2005)

SHOW-ME DOCTORAL FELLOWS
Jennifer Bay-Williams (1997–1999)
Ann Bledsoe (1997–2002)
Deanna Wasman (1997–2000)
Ken Simms (1997–1998)
Bob Glasgow (1998–2000)
Gay Ragan (1998–2000)
Mark Taylor (1998–2001)
Brian Townsend (2001–2005)
Dustin Jones (2001–2004)
Bridgette Stevens (2001–2002)
David Barker (2001–2004)

ADVISORY BOARD
Hyman Bass, University of Michigan
Diane Briars, Pittsburgh Public Schools
Mary Lindquist, Columbus State University
Sid Rachlin, East Carolina University
Linda Rosen, Consultant
Ed Silver, University of Pittsburgh/Michigan
Jane Swafford, Illinois State University

PROJECT EVALUATORS
Illinois State University
Jane Swafford (1997–2002)
Cindy Langrall (1997–2007)
Edward Moody (2002–2007)

SATELLITE STAFF
Connected Mathematics Project Satellite
Michigan State University
Betty Phillips, Satellite Director (1997–2007)
Glenda Lappan (1997–2007)
Mathematics in Context Satellite
University of Wisconsin
Meg Meyer, Satellite Director (1997–2007)
Jack Burrill (1999–2001)
David Webb (2002–2007)
MathScape Satellite
Educational Development Center
Susan Jansen, Satellite Director (1997–2002)
Emily Fagan, Satellite Director (2002–2005)
Fred Gross, Satellite Director (2005–2007)
MATH *Thematics* Satellite
University of Montana
Rick Billstein, Satellite Director (1997–2007)
Jim Williamson (1997–2007)
Mathematics Modeling and Applications (MMAP) Satellite
Institute for Research on Learning
Jennifer Knudson, Satellite Director (1997–2002)
Shelly Goldman (1997–2002)

REGIONAL ASSOCIATES
Robert Bates, MA
Jenny Bay-Williams, KS
Peg Bondorew, MA
Murrel Brewer-Hoover, WV
Judd Freeman, CO
Rebecca McGraw, AZ
Nancy Mumaw, OH
Ron Preston, NC
Gay Ragan, MO
Linda Selwood, DE
Sharon Stenglein, MN

SHOW-ME MASTER TEACHERS
Linda Tetley, Jefferson City MO Schools
Yvonne Grant, Portage MI Schools
Amy Dougherty, Waltham, MA Schools
Margie Pliggie, Horicon, WI Schools

work of the project. While the curriculum development representatives provided input into all major decisions, much of the work of carrying out the decisions rested on the Missouri team. This team, under the direction of Barbara Reys, planned and carried out the major activities of the project including national and state curriculum showcases, researchers workshops, teacher education conferences, Show-Me website development, and promotion of the work of the project via published journal articles. In addition, each curriculum development team took leadership responsibilities for one major joint initiative:

- MT Satellite Center (University of Montana)—disseminating information to teacher educators.
- CMP Satellite Center (Michigan State University)—development of a manual to facilitate professional development in support of standards-based curriculum implementation.
- MS Satellite Center (Education Development Center)—development of materials to support curriculum implementation with special needs students.
- MiC Satellite Center (University of Wisconsin)—collection of resources related to assessment aligned with standards-based curriculum.
- MMAP Satellite Center (Institute for Research on Learning)—development of materials to assist parents in understanding standards-based curriculum.

As the first round of funding for the Show-Me Project came to an end, the Missouri staff made the decision to invite the four comprehensive middle school curriculum development projects (MT, CMP, MS, and MiC)[2] to work together to submit a renewal proposal to NSF. The proposed 5-year renewal was positively reviewed by NSF but funded for only an additional 3 years (see Chapter 2 by Spud Bradley for this story). The major activities of the first 8 years of the Show-Me Project are described in the following sections.

National Middle School Curriculum Showcases

From 1998 to 2006, the Show-Me Project sponsored ten national conferences. Seven conferences showcased the curricula, one conference targeted teacher education, and two were workshops for curriculum researchers. In addition, two targeted one-day conferences for teacher educators were held as pre-sessions to the annual conference of the Association of Mathematics

Teacher Educators. The curriculum showcase conferences addressed the following goals: (a) to introduce teachers and teacher leaders to the NSF-funded middle grades mathematics curricula, (b) to support current users of the NSF-funded middle grades mathematics curricula through workshops by experienced users, and (c) to provide forums for exchanging information and considering issues related to implementation of the NSF-funded middle grades mathematics curricula. Although the Center and Satellite directors made changes in conference sessions every year in response to participants' feedback, the format of the conference was generally the same. Participants were able to choose to attend a variety of breakout sessions pertaining to each of the curricula. These sessions typically addressed specific content strands (e.g., rational numbers, algebra) and implementation strategies or issues (e.g., effective teaching strategies, development of student thinking, drawing out the important mathematics). General breakout sessions, as well as keynote talks that were not curriculum specific, were also offered. Topics for the general breakout sessions included the following: assessment, parental outreach, technology, teacher panel discussions, and other projects related to curriculum implementation.

Publishers provided sample copies of their middle school mathematics materials to each participant at the Show-Me conferences, thus allowing participants to take home actual student and teacher materials to use in their schools and share with other teacher leaders. Publishers also provided refreshments for the conferences on a rotating basis. Overall, the publishers made significant contributions to the success of the Show-Me conferences and in return they made important contacts with schools interested in adopting their curricula.

The Show-Me Curriculum Showcase conferences were remarkably successful in accomplishing their goals. On average the conferences hosted about 300 participants per year, many representing their school districts as well as state and national curriculum leaders. Participants reported that the information and materials they received at a conference were shared with school district administrators, teachers and other colleagues, and in work with preservice teachers. Participants indicated that their attendance at a conference directly or indirectly influenced a school district's textbook adoption process. Participants also perceived the conferences as effective forums for examining issues pertaining to the implementation process and for promoting standards-based mathematics instruction at the middle grades.

Show-Me Website

The Show-Me website (http://www.showmecenter.missouri.edu) continues[3] to serve as a key tool to disseminate information relative to the

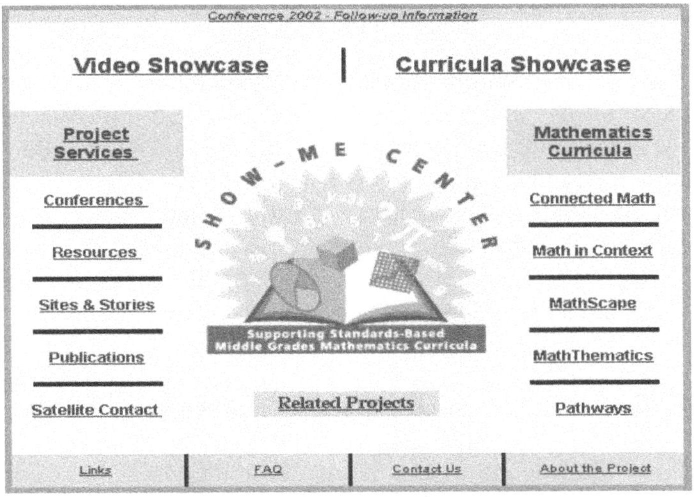

Figure 3.2 Show-Me Project website home page.

NSF-funded middle grades mathematics curricula. The website came online in November 1997 with regular additions of new features and information throughout the 10-year period of the grant. The site contains information about the Show-Me Project, curriculum showcases for each of the curricula, mini web pages for each curriculum with links to publisher and developer websites, a database of workshop opportunities and conference presentations, access to published articles related to standards-based curricula, answers to frequently asked questions, links to related websites, and a mechanism to request information (see Figure 3.2 for the website home page). In addition to a "Curriculum Showcase" which includes a searchable database of online lessons from the curricula, a "Video Showcase" is also available. The video showcase includes video clips of lessons taught by teachers using the curriculum materials. This resource was developed in collaboration with the *Modeling Middle School Mathematics* Project funded by NSF (Bolster & Reys, 2002). Throughout the period 1998–2003, the site regularly registered about 100,000 hits per month. The Center conducted a review of the website in 2001 using an on-line survey to gather feedback from 20 mathematics educators. The results from the survey were very positive, noting a well-organized site containing useful information. In addition to the Center website, the Satellites and/or their publishers maintain their own websites, which contribute further to the achievement of the dissemination goal.

Curriculum Researcher Workshops

Two Show-Me sponsored "Researchers' Workshops" were held in 2003 and 2004. About 50 advanced graduate students and junior faculty members at each workshop participated in sessions designed around four working groups: Curriculum Analysis, Instructional Materials & Teaching, Instructional Materials & Teachers, and Instructional Materials & Student Learning. The goal of each workshop was to stimulate more research efforts related to the impact of curriculum materials on teacher knowledge and practice and student learning.

Summer Curriculum-Specific User's Conferences

The Curriculum Satellite Centers each hosted a "users" conference each summer to focus on their particular curriculum materials. These ranged in size and generally included school district curriculum leaders and/or lead teachers (for more detailed information about these User Conferences, see other chapters in this book).

Publications

Show-Me staff published over 40 articles (see these and others at: http://showmecenter.missouri.edu/showme/pp.html) during the first round of funding (1997–2002) and continued publishing a wide range of papers as part of the dissemination mission. In addition, the staff proposed and edited a feature ("Innovations in Curriculum") in the NCTM journal, *Mathematics Teaching in the Middle School* from 1998–2000. Eight articles appeared during 1998–2000, four by teachers reflecting on their experience using standards-based curricula.

The staff also produced materials to support curriculum implementation including a *Professional Development Handbook, Research Brief on Curriculum and Student Learning,* and a set of videos illustrating teachers' use of the curricula (see http://www.mmmproject.org/homeS.htm).

The purpose of the Show-Me Center was to promote the dissemination of the NSF-funded middle school mathematics curricula. In all of its work (e.g., Conferences, website, and presentations), the Show-Me staff provided non-evaluative information about the different NSF-funded middle school mathematics curricula, their use, and impact on teachers and students. Consequently, the Show-Me staff never made evaluative judgments, and despite many requests to do so, the Show-Me Staff refrained from recommending one curriculum over others.

Collaboration with Other Projects

In addition to providing a structure for continued interaction and collaboration among the five curriculum development teams, the Show-Me Project also served as a focal point for collaboration with other projects. In fact, the Show-Me Project had formal connections with at least five separate projects (see Table 3.2).

Collaboration across Curriculum Development Projects

Collaboration of author teams across competing textbook series is unusual. The nature of the competitive school market makes it unlikely that such collaboration will be stimulated naturally. That is one of the unique contributions of the four Implementation and Dissemination Centers. Due to their mission, organizational structure, and budget, different author teams were regularly brought together to collaborate on a variety of initiatives. These collaborations were a direct result of ARC, Show-Me, COMPASS, and K–12 MCC structure and initiatives that they launched.

TABLE 3.2 Projects in Partnership with the Show-Me Project

Project	Principal Investigator	Description
Modeling Middle School Mathematics (MMM)	Robert Reys Cary Bolster	Produced two videos for each middle school mathematics curriculum. Each video showed teachers using the curriculum in their classes. For more information see www.mmmproject.org
$(CM)^2$	Ira Papick	Developed a series of mathematics textbooks for use by preservice teachers using lessons from the curricula. See the chapter by Ira Papick in this book.
TinkerPlots	Cliff Konold	Developed statistics software for use by middle school students. Satellite representatives incorporated *TinkerPlots* into revised versions of their curricula. For more information see www.keycurriculum.com
COMET	Ed Silver	Satellite staff provided problems for use in this project.
ESCOT	Andy Zucker, Ed Esty	Developed a software program *Math Workshop* for use by middle school students. Satellite representatives served on the advisory board.

Collaboration with Publishers

A unique and important aspect of the work of the Show-Me Project was its collaboration with commercial publishers. As noted earlier, each curriculum development team sought a contract with a textbook publisher as a requirement for continued funding from NSF. In addition, each of the publishers agreed to support the work of the Show-Me Project through collaborations and contributions. For example, each publisher hosted one of the Project's National Advisory Board meetings. In addition, each publisher donated materials to the library at the University of Missouri, to the other Satellite Centers, and to participants of the Show-Me sponsored curriculum showcases. The different publishers understood the uniqueness of the activities conducted by the Satellite Centers and recognized the value of working together to support dissemination and implementation efforts under the Show-Me umbrella. The value of this partnership was undeniable. However, its potential would likely have been greater had there not been such a high turnover rate among publisher personnel. The high frequency of personnel changes by the publishers (including editors and sales representatives) resulted in new publisher people working with the Show-Me Project but having no knowledge of previous work.

The Math Wars

One of the unexpected challenges that surfaced as the NSF mathematics curricula were being published and adopted was opposition to curriculum reform efforts from different quarters. Often, parents did not understand the changing mathematics curricula their children were experiencing and raised questions. These questions were sometimes curriculum specific and needed to be addressed at the local level. Some mathematicians also expressed concerns about the mathematics content of the new curriculum materials. Some criticism was leveled at the way certain topics were developed while other concerns focused on the lack of attention to procedural skills. While most of these concerns were voiced at the local level as school districts were reviewing new curriculum materials, the complaints gathered attention at the national level when small, organized groups of individuals began exchanging information and posting discussions on websites, such as "mathematicallycorrect.com."

Further discontent was stimulated by a review of curriculum materials sponsored by the U.S. Department of Education. In 1999 an independent review panel commissioned by the Department of Education identified "exemplary" and "promising" mathematics curriculum materials, and several NSF-supported curricula led the field of highly-rated materials. The

Department of Education sponsored review was criticized by a group of prominent research mathematicians in a full-page *Washington Post* ad and launched what came to be called the "Math Wars." The Math Wars were not directed exclusively at middle school mathematics, but often took the form of a broad attack on all of the NSF-funded mathematics curricula. A number of websites and articles were developed to address critical issues and many of these revealed organized opposition to reform-based efforts initiated by NCTM and/or NSF. It also sparked many articles directed to lay audiences (Becker & Jacob, 2000a, 2000b; Tepper-Haimo & Milgram, 2000) as well as professional mathematicians (Kilpatrick, 1997; Wu, 1997). These debates sparked additional articles calling for reasoned discussions, along with efforts to establish common ground continued for more than a decade (Reys, 2001b; Schoenfeld, 2004; Ball, Ferrini-Mundy, Kilpatrick, Milgram, Schmid, & Schaar, 2005).

The Show-Me Project's response to the Math Wars was to provide resources in the form of information to guide local school districts in responding to questions about the philosophy and content of the NSF-funded curriculum materials. Information included research briefs, FAQs, and articles in newspapers and professional journals (for example, see Reys, 2001a). The Show-Me Project also became more active in stimulating researchers to engage in studies to document the impact of the new curriculum models. For example, the project sponsored two "Researcher' Workshops" that engaged young researchers in sharing instruments and research designs. The Math Wars environment also prompted the project to identify and enlist a group of "Show-Me Regional Associates" to help local districts address issues raised by opponents of local curriculum reform efforts. These regionally-based professionals were experienced teachers and curriculum leaders skilled in working with parent groups and school boards. Finally, the four mathematics Dissemination and Implementation centers collaborated in hiring a public relations specialist (Marianne Smith) to consult on ways the Centers could provide a balanced response to the critiques of curriculum reform efforts.

CLOSING COMMENTS

Collaborative efforts are powerful in stimulating and supporting change, and the establishment of national centers, such as the Show-Me Project, provided opportunities for multiple mathematics curriculum projects to inform, share and demonstrate their materials to teachers/leaders throughout the country. The Show-Me Project provided an efficient and powerful means of disseminating information on all of the comprehensive NSF-funded middle school mathematics curricula via conferences and the website.

The collaborative efforts through the Show-Me Project resulted in the sum of the efforts from the mathematics curriculum projects exceeding what could have been done by them operating individually. Furthermore, the organizational structure made it easier for the Show-Me Project to serve as a backbone for related projects, such as M^3 and $(CM)^2$ as these projects worked closely with all of the middle school mathematics curricula. It also allowed for the Show-Me Project to work together with the high-school implementation center (COMPASS), elementary center (ARC), and K–12 center (K–12 MCC) on issues and topics of mutual interest (Reys, Robinson, Sconiers & Mark,1999).

The last solicitation for Dissemination and Implementation Centers was in 2001 and resulted in a 3-year renewal of the Show-Me Project. None of the NSF solicitations since 2001 included calls for Centers. Instead, in 1999 NSF initiated a call for Centers for Learning and Teaching (CLT). These Centers differed from IMD Centers in that they focused on building the capacity of the field to engage in leadership and research. Specifically, the goals of the CLT initiative are to renew and diversify the cadre of leaders in STEM education; to increase the number of K–16 educators capable of delivering high-quality STEM instruction and assessment; and to conduct research into STEM education issues of national import (e.g., the nature of learning, teaching strategies, and reform policies and outcomes). As a result of the Show-Me Project's efforts to build a network of mathematics curriculum scholars, some of the staff submitted a successful NSF proposal to launch the Center for the Study of Mathematics Curriculum (CSMC, 2004–08). CSMC serves the K–12 educational community by focusing scholarly inquiry and professional development on issues of mathematics curriculum. Major areas of work include understanding the influence and potential of mathematics curriculum materials, enabling teacher learning through curriculum material investigation and implementation, and building capacity for developing, implementing, and studying the impact of mathematics curriculum materials. In addition, CSMC supports the development of leadership capacity (particularly with regard to doctorates in mathematics education with expertise in curriculum design, analysis, and evaluation).

CSMC represents the current effort to support the development, implementation, and evaluation of mathematics curriculum. In particular, the aim is to support the development of the next generation of curriculum developers, school-based curriculum leaders and curriculum researchers in the field. CSMC has its roots in the work of the Show-Me Project and represents what can be accomplished by a continued line of support from the NSF.

NOTES

1. The Show-Me Project initially included the Show-Me Center at the University of Missouri and five Curriculum Satellite Centers at University of Wisconsin, University of Montana, Michigan State University, Educational Development Center and the Institute for Research on Learning.
2. The MMAP program did not develop into a stand-alone comprehensive curriculum. It was designed as a technology-intensive mathematics curriculum and was most commonly used to supplement other materials.
3. This website was still available at the time of publication of this book.

PART II

ADOPTION CASE STUDIES AND IMPLEMENTATION STORIES

This part of the book is organized in two themes: *Issues of Curriculum Adoption* and *Contexts for Implementation.* The first four chapters focus on curriculum adoption and offer different perspectives on the process of reviewing and adopting curriculum materials. They highlight why this process is different for standards-based materials than for traditional curriculum materials. The next four chapters present curriculum-specific implementation stories that report the successes and challenges of implementing NSF-funded mathematics curricula. These are followed by a more general chapter that presents a set of four "hypothetical" cases, designed to illustrate some of the particularly difficult challenges and barriers districts have faced. The final chapter shifts the focus to two specific implementation issues related to language/literacy and mathematics for *all.* The authors provide recommendations based on research and grounded in current related literature.

The following *Reflection Questions* can be used to organize individual reading of these chapters as well as for group discussion.

CURRICULUM ADOPTION/IMPLEMENTATION QUESTIONS

1. What challenges did these districts face as they started the process of adopting (implementing) an NSF-funded mathematics curriculum? Is there anything in common across the four stories in this regard? What did each district do to meet their particular challenges?

2. What seemed to motivate each district to begin the process of adopting a new curriculum? Why were they particularly interested in the NSF-funded curricula?
3. What was the decision-making process in each adoption case study? How were various stakeholders involved?
4. What "outside" influences (either stated or implied) did each district encounter during the adoption (implementation) process?
5. What systemic changes did each district make in order to implement their curriculum? What evidence did they gather to evaluate the effectiveness of their implementation?
6. What role did professional development play in the implementation for each of these districts? What evidence is there that the districts had long-term plans for professional development?
7. If you could ask these authors questions, what else would you like to ask about their curriculum adoption (implementation) experiences?
8. What "lessons learned" do you take away from these four adoption (implementation) stories?
9. How do these stories reflect your own district adoption (implementation) situation?

CHAPTER 4

TEACHER INVESTMENT IN CURRICULUM ADOPTION

The Role of Piloting Materials

Cary Tuckey
Kirkwood R-7 School District (retired), Kirkwood Missouri

In 1995, the Kirkwood School District was nearing a mathematics adoption year when we received an invitation to be part of the Missouri Middle School Mathematics (M^3) Project at the University of Missouri–Columbia. This ended up being a professional development experience like no other. It helped us support our eventual decision, brought two middle school mathematics departments to the same page philosophically, and gave us the knowledge and research we needed to convince our school board, administrators, and parents of the need for a major shift in the way we taught mathematics in the district. The philosophy of the M^3 Project was that important changes in middle school mathematics programs do not simply happen; nor can they be mandated or pushed upon unwilling participants. Rather, the project aimed to empower teachers to make informed decisions about curriculum and instruction in their classrooms. With support from our school district to seriously reconsider our middle school mathematics curriculum, provide resources to field test materials, and pay substitutes for

A Decade of Middle-School Mathematics Curriculum Implementation, pages 35–41

release days for all mathematics teachers involved, we joined the three-year project. By the end of our second year of participation in the M^3 project, we had adopted the *Middle School* MATH*Thematics* program (Billstein & Williamson, 1999a). Our third year of participation in the project supported the implementation of these materials.

Even though Kirkwood is a small district with only two middle schools (approximately 5000 students), our teaching staff had a wide range of experiences and philosophical views about mathematics teaching and learning: we had very traditional teachers, teachers who had embraced a more investigative approach to instruction, teachers new to the profession, and veteran teachers. Going into M^3 we were not philosophically aligned. We all "did our own thing" and shared what we did, but there was not much consistency. Of course, unanimous agreement on anything in education is difficult to achieve.

PROFESSIONAL DEVELOPMENT INFORMS THE ADOPTION PROCESS

The first M^3 meeting focused on learning about reform mathematics and general features of the NSF-funded mathematics programs. We heard about the whys of changing mathematics instruction, we looked at issues, concerns, and questions related to middle school curriculum, and we focused on what teachers could do to change and improve middle level mathematics. Often one or two people attend a great conference, but when they come back to share what they have learned, the impact is diminished. The fact that almost all teachers from Kirkwood had the same professional development experience, I think, made the difference. With only two or three of our teachers not attending M^3 sessions, it was easy for our monthly department meetings and quarterly joint department meetings to be filled with enthusiasm and discussions that were engaging and thought provoking for everyone. The experience also brought together the mathematics departments from our two middle schools and created friendships as well as collegiality of which other departments were literally envious.

Piloting Materials

We spent the balance of the first year reviewing each NSF-funded program. Three- or four-day sessions were conducted by the actual author teams. The philosophy of each program was highlighted, and we were literally taken through units and activities as if we were the students. The presenters modeled the same instructional style that a teacher would use in the classroom. After each experience, we took units/modules back to our classrooms and taught them. Following each M^3 meeting, we shared our experiences, questions, and opinions with the teachers who were not involved in the project. We reviewed each of the curriculum programs so that the teachers who did not attend the M^3 sessions were also able to pilot each program in their classrooms.

As we began to pilot materials, we soon realized that time and pacing would be an issue. Because this was a new experience, we spent more time teaching each lesson than anticipated. We did not want to cut any unit/module short, so we continued until each was completed. Nevertheless, we were expected to cover our district's curriculum and so we chose to pilot a different content strand for each set of materials. It was beneficial for all of us at a grade level to be piloting the same materials at the same time so we could have conversations along the way. During previous adoptions we had been encouraged to pilot new materials, but textbook companies did not give each teacher a set of books, or even a common unit, to use. Thus, each teacher was often piloting a different program. Although we had forms to complete about the characteristics of each book, I know I always leaned more heavily toward the one I had piloted because I was most familiar with it. Without actually piloting a program, it is difficult to get a sense of how the approach will work with students.

As a teacher, piloting gave me a chance to see how the lessons were presented, how the manipulatives were used, and how students were encouraged to investigate the problems posed. Also, we all had the chance to experience firsthand the shift toward teaching a standards-based curriculum. Because of the excitement of our kids, most of us found that the adjustments to our instructional practice were well worth the effort. Through our follow-up discussions after each pilot it became evident that we, too, were excited about the knowledge we had gained, the experiences we had had, and the real possibility that we could adopt one of the NSF-funded curriculum programs.

After the pilot of MATH*Thematics*, which was our third, we focused on how topics could be sequenced as a spiral so that they were revisited within each grade level, as well as grade to grade. We talked about the use of "replacement units" versus a sequenced curriculum that was embedded in themes rather than mathematical content strands. For example, if a teacher did not like or was not comfortable with a particular unit of instruction, it would be very easy to skip that unit, fall back on old favorites, and then pick up with the next unit. In order to have some standard of instruction to which our teachers were held, we felt that a curriculum that spiraled within the year would "force everyone's hand." It is not possible to just pull all of the fraction explorations out of MATH*Thematics* and teach them as a fraction unit. Because the themes change throughout the book, the individual lessons or activities would not make sense together or create a cohesive unit of study. We also saw this as an advantage for teachers who did not have the full M^3 experience. Using MATH*Thematics*, teachers in the district would have something solid to hang on to as a basis for instruction. After all, as a district we had adopted a philosophy of standards-based mathematics instruction and it was important for all teachers, present and future, to buy into that philosophy. By the end of our first year in the M^3 project, we had piloted all four of the NSF-funded curriculum projects, and it was becoming evident that we were leaning toward adopting MATH*Thematics.*

A Decision to Adopt

The Kirkwood School District has a formal protocol for textbook adoption and implementation. The process is data driven and involves an evaluation of student performance and review of the district curriculum framework as precursors to adopting and implementing curriculum materials. The cycle is characterized as follows:

- Analyze student performance data
- Research best practices
- Revise district curriculum framework
- Adopt curriculum materials that align and support the district framework
- Provide professional development for implementing curriculum materials
- Develop and pilot assessments to collect data on student performance

Our continued involvement in the M^3 project supported us in this process. We examined standardized tests and the content they assessed, analyzed our students' achievement data, learned about mathematics education reform, reviewed the district curriculum framework, and shared our philosophies about teaching mathematics (which were changing due to our participation in M^3). Through our piloting the four NSF-funded curriculum projects, we were able to determine which one best fit our needs. We knew we were ready to move from a traditional way of teaching strand-by-strand to something that would allow our students to experience all strands of mathematics all year long. We did not want a project whose units/modules could be used as replacement units.

We discussed the pros and cons of each program and agreed that we wanted a common way to address the curriculum within the district. MATH*Thematics* was the only NSF-funded project that fit our curricular and philosophical needs. Thus in Year 2 of the M^3 project, we examined MATH*Thematics* in more depth, piloted more of the materials, and cross-referenced our district mathematics framework with the MATH*Thematics* materials to verify alignment between the two. We continued to attend M^3 meetings, which provided more opportunities for planning, discussion, and reflection, in addition to sessions on the use of graphing calculators, performance-based assessment, and the use of manipulatives and models to support instruction.

It is fair to say that all middle school mathematics teachers in Kirkwood had an equal say in the adoption process. The decision to adopt MATH-*Thematics* was definitely made from the bottom up. With few exceptions, all middle school mathematics teachers participated in the M^3 project. The teachers who could not participate were always included in piloting the materials and were part of the discussion and evaluation process. We had discussions within our own buildings about each curriculum project, and when we came back together in district-wide meetings, everyone had an equal voice in the decision making. The fact that we were committed to investing the time together was a major reason the process went so smoothly and everyone was satisfied with the outcome.

Some often wonder why teachers balk when given new curriculum materials and why they are dissatisfied with adoptions. Although it is a time-consuming process, having teachers who are going to be using the materials involved from the start empowers teachers by encouraging them to participate in the process and make their voices heard. If teachers feel their concerns are addressed, their positive attitude will carry over to classroom instruction and provide students a better experience. Since teachers are responsible for creating a positive learning environment for students, why wouldn't a district want to invest the time to allow the teachers to be involved?

Another issue any district faces when new curriculum materials require a change in instructional style and strategies is making sure that administrators, parents, and board members are informed. Our district mathematics facilitator was involved in the process from beginning to end. The principals at both middle schools were kept up-to-date on our experiences during the piloting process and visited our classrooms to observe what was happening. The central office administration was also kept informed along the way. Once a decision had been reached, the district facilitator, a teacher from the other middle school, and I made a presentation to the school board. We shared actual samples of student work, which amazed those who attended. They were impressed that sixth graders could successfully handle the types of problems that were being put before them. After the presentation to the school board, the adoption of MATH*Thematics* was approved.

LESSONS LEARNED

As with any change, there are always obstacles and challenges to be faced. As a middle school mathematics staff, we had to really wrestle with our philosophy regarding mathematics instruction. Many of us had been teaching for a while and had taught using traditional strategies. Some of us had switched to more investigative strategies after attending a Marilyn Burns workshop, and even though we still used traditional textbooks, we created our own units of instruction using a more constructivist approach. However, through our participation in the M^3 project, we came to realize the importance of a coherent instructional program to which all teachers would be held accountable. We believe that MATH*Thematics*, with its spiral approach that revisits topics from unit to unit and across grade levels, will help us maintain such a program. Although it took a lot of time and energy for teachers to agree upon a similar philosophy of teaching and learning, this process was necessary to bring mathematics instruction in our district to a more consistent level. It may be of interest to note that Kirkwood completed another adoption cycle during the 2005–2006 school year and, because of our satisfaction with the way mathematics is being taught and learned, chose to stay with MATH*Thematics*.

Although our experience with the M^3 project may not be available to most districts, there are several aspects of our adoption process that can be generalized to any district:

- Take the time to understand why changes are needed in the way mathematics is typically taught. Take the time to get everyone on board, especially if this is a major shift in philosophy. See what the research says, and more importantly, have serious philosophical discussions involving as many people as possible. Even though it may be painful, try to come to consensus.
- Have as many teachers pilot materials as possible. Don't just flip through the books and fill out charts for feedback. Get together and have in-depth conversations about teachers' experiences, the format of the books, resources, and students' reactions.
- Get parents and administrators involved from the start. Invite parents and administrators to observe in classrooms when materials are being piloted, understanding that this is the first time that teachers and students are using the material. Then talk with them. What did they see? Teachers can gain a lot of insight when they "see" their classroom from another perspective.
- Insist that the publisher provide whatever professional development you think is necessary. It is invaluable to have teachers who have actually taught with the materials conduct the inservice. They can

tell you what really happens in the classroom with respect to management, manipulatives, grading, differentiating instruction, and so on. If you are not satisfied with the level of professional development the publisher provides, ask your district to bring a teacher in for a day or two. The better prepared teachers are to make a shift in instructional style, the better off students will be.

- Observe teachers who are using the materials in which you are interested. See for yourself what happens in the classroom. Sales representatives should be able to help you make contact with someone else in your area who is using the same materials.

Our in-depth involvement with the M^3 project helped us understand why mathematics instruction needs to change in the United States. Through this project we became familiar with the NCTM *Standards* (1989) and heard many experts speak about research, methodology, and philosophy supporting the development and implementation of curriculum materials. Having three years of professional development together allowed us to develop a shared vision for the teaching and learning of mathematics. Most importantly, we talked, and talked, and continue to talk with each other. Although the process was time consuming, if all districts could try to simulate an experience such as this, I think adoption decisions would be much more successful.

CHAPTER 5

CONSENSUS BUILDING

A Key Component of Curriculum Adoption

Cheryl Rectanus
Portland Public Schools, Portland, Oregon

In early 1994, the superintendent of Portland Public Schools (PPS), Oregon, agreed that a "summit" of teachers, administrators, parents, and community members be brought together to make recommendations for improving achievement in mathematics and science district wide. This summit of over 300 people set the course for the next five years with recommendations for increased requirements in mathematics and science and improvements in instruction. Committees of teachers and administrators met to design an implementation plan to achieve these goals. Prior to the summit, significant work had taken place that created the conditions that led to elementary, middle, and high school mathematics adoptions in 1998–1999.

Starting in the late 1980s, many PPS K–12 mathematics teachers had attended annual PPS summer institutes that focused on improving both their content and pedagogical knowledge of mathematics. The institutes helped teachers understand the changes being called for in the 1989 NCTM *Standards*, which created fertile ground for discussions about the kinds of in-

A Decade of Middle-School Mathematics Curriculum Implementation, pages 43–54

structional materials that would support these changes. In the early 1990s, several K–12 teachers piloted some of the NSF-funded curricula, in particular *Connected Mathematics* [*CMP*] (Lappan, Fey, Fitzgerald, Friel, & Phillips, 1998), *Investigations in Number, Data, and Space* (Russell, Tierney, Mokros, & Goodrow, 1997), *Math Trailblazers* (Wagreich, Goldberg, Beiler, & Simpson Beissinger, 1997), COMAP (1998), and *Core-Plus Mathematics* (Coxford, Fey, Hirsch, Schoen, Burrill, Hart, & Watkins, 1998). The teachers who piloted *CMP* were so excited that at a summer mathematics institute they shared what they learned, prompting three middle schools (two characterized as low socioeconomic status) to pilot *CMP* school-wide in the mid-to-late 1990s. At that point, conditions were ideal for seriously considering a district mathematics adoption. PPS's curriculum department included a team of Teachers on Special Assignment (TOSA). They provided the infrastructure necessary to support high-quality mathematics teaching and learning in PPS schools. Four TOSAs worked at the elementary level with 64 schools, two worked at the middle school supporting 19 schools and a handful of special programs, and one high school TOSA had responsibility for addressing the needs of our 12 high schools. While I am now a secondary mathematics specialist and coordinate the middle school mathematics program for PPS, at that time, I served on the middle school textbook adoption committee as a teacher representing one of the middle schools.

MATHEMATICS TEXTBOOK ADOPTION COMMITTEES

During September of the 1998–1999 school year, district-wide Mathematics Textbook Adoption Committees were convened at the elementary (pre-K–5), middle (6–8), and high school (9–12) levels. They were charged with reviewing and recommending mathematics materials for adoption.

The Middle School Mathematics Textbook Adoption Committee was co-chaired by the Middle School TOSAs. In order to represent the needs and concerns of all the teachers in our large district, the committee was comprised of one teacher from each of the middle schools in the district, as well as a representative from the Special Education and English as a Second Language (ESL) departments, two administrators, one union representative, and the K–12 Mathematics Coordinator. Committee members were selected based on their experience in teaching or supporting middle school mathematics, their knowledge of current issues in mathematics education, and the leadership they provided in their schools. They acted as liaisons between the committee and their school, keeping the mathematics teachers informed of the textbook adoption process and soliciting their input throughout all stages of the process.

The district did not have a standard procedure for curriculum review and adoption, so the adoption committee's first task was to develop a vision statement for mathematics in PPS middle schools. To do this, the committee read articles and reviewed various curriculum summaries, research on effective instructional practice, and the NCTM *Standards* (1989). We asked our building colleagues to identify what was important to them in a textbook or program, and elicited their input about criteria we should use in evaluating instructional materials. The challenge was to represent the views of all of the mathematics teachers in our schools, especially since at times members of the adoption committee did not agree with those views. It was also difficult because in some schools, the mathematics teachers were not particularly interested in the adoption process. In others, teachers were very concerned. They had definite opinions about the kind of curriculum they thought would best serve their students and thought that the committee should simply get down to the business of selecting the "best" curriculum materials, which were, not surprisingly, the ones for which they were advocating. We spent time with these teachers to explain that the process affected more students than those in their schools and why we needed to have a process in place that we could all support. Such conversations were often not easy.

In all three adoption committees, commonalities emerged. For K–12 mathematics, we wanted a program that:

- Was coherent, relevant, and tied to state and national standards
- Enabled all students, not just those currently successful, to demonstrate proficiency in mathematics
- Emphasized mathematical reasoning, problem solving, strong number sense, and communication
- Addressed a broad range of mathematics
- Supported a variety of instructional formats and strategies

These recommendations were shaped by the beliefs that:

- No curriculum will improve achievement by itself, but a curriculum that meets the criteria supported by improved teaching will enable schools to achieve these goals.
- Professional development and time to collaborate within and between schools is key to improvement.
- Technology should be used to its full potential both in teaching and in professional improvement.

All three committees used this vision statement as a basis for constructing an evaluation form for reviewing materials. Each adoption committee wrote specific criteria and developed an evaluation scale for their respective grade levels (see Appendix). However, while the TOSAs regularly discussed the progress of their respective committees, the three groups did not collaborate with one another or look closely at the articulation between the elementary and middle materials or the middle and high school materials. In retrospect, it would have been useful to collectively analyze the implications of our recommendations across grades K–12, especially if the pedagogical approaches of the materials we were considering were vastly different.

EVALUATION OF INSTRUCTIONAL MATERIALS

In 1998, there were 11 middle school programs on the Oregon State textbook adoption list including four NSF-funded programs. We decided that evaluating all 11 was the only way to be sure that we were not inadvertently overlooking a set of materials that might fit our criteria. We found it relatively straightforward to evaluate the traditional mathematics textbooks since they were familiar to all of us on the committee. We understood their structure, organization, advantages, and disadvantages because most teachers had used them. It was harder to evaluate the NSF-funded curricula (i.e., *CMP*, *MathScape*, MATH*Thematics*, and *Mathematics in Context*) by reading the texts and teacher guides. When we opened up a teacher's guide, it looked different than a traditional textbook. It was not clear how to navigate through the guide and we had to spend a lot of time reading the materials to understand what they were about. It was also clear that these programs were different from what many teachers were used to teaching. Consequently, we decided to pilot all four programs to learn about their instructional approaches and how they might impact student learning.

In early fall, the committee agreed that some of the members of the committee would pilot one unit from each of the four NSF-funded curricula in order to inform the discussions that continued throughout the year. Three or four teachers at the 6th, 7th, and 8th grades volunteered to "give up" some of the curriculum they were teaching in order to take on this task. They were not teachers who were already piloting *CMP*. The committee-at-large thought that it made the most sense to look at four units selected from the four programs focused on number, algebra, statistics, or geometry so each content strand was adequately represented. We developed a feedback form for the pilot teachers to record evidence of how the unit met the evaluation criteria. Teachers were also asked to give specific evidence for each of the criteria, to record general strengths and weaknesses of the program, and to report the reactions students had to the program (see Appendix).

As we moved through the process, issues emerged. Should we adopt a single series or allow individual schools a choice? After much discussion, an important point of consensus emerged: teachers overwhelmingly recommended adopting a single set of materials to address the high levels of student mobility in our district and to focus professional development efforts. This was a huge change for our large district where teachers were used to doing what made sense to them regardless of what the teacher next door or across the city was doing. This led to another important discussion: there was no such thing as a "perfect" curriculum and no matter what we adopted, its use would need to be tailored to the specific circumstances of our district. As part of this discussion, we wondered if it was easier to infuse skills practice into a problem-based curriculum or to infuse problem solving into a skills-based curriculum. We felt fairly confident about supplementing with skills, but not problem solving. By early spring, the committee had narrowed the original 11 textbooks down to 4, three NSF-funded curricula and one traditional program. Now we needed a more careful look at the materials.

To begin the process of reviewing the final four sets of materials, we broke into small groups that focused on a different content area (proportional reasoning, algebra, and geometry). We looked at how specific topics (e.g., slope) developed throughout the materials over three years, from unit to unit in the NSF-funded curricula or chapter to chapter in the traditional textbook. We carefully examined how cohesive the programs were. Tracing specific concepts helped us make this determination. In doing this we found that two of the programs were less coherent than they had initially appeared.

In March, we met and shared our findings. The committee overwhelmingly agreed that the traditional program and one of the NSF-funded programs did not meet our criteria. But the remaining two NSF-funded curricula did. Now the challenge was to figure out a way to make a single recommendation.

We heard from the teachers who piloted each of the two sets of materials and from those who had evaluated the materials. While a few units from each of the curricula were piloted, aggregating the information was necessary to see the big picture of each program. The teachers who represented schools piloting *CMP* shared their colleagues' experiences with the curriculum. We examined assessment data from schools that had used *CMP* for several years prior to the adoption year and noted the significant improvement in test scores, especially in less affluent schools. We also examined the ratings of each curriculum by the American Association for the Advancement of Science (AAAS, 2000). In the end, it was the assessment data and each program's coherence that most strongly influenced our recommendation.

RECOMMENDATION

After much debate, we recommended the adoption of the *CMP* materials. One part of the design of the decision-making process was that the decision had to be a consensus and everyone on the committee had to agree not to sabotage the program selected. This was new territory for most of the committee members who were used to a fair amount of autonomy, yet the group realized why this was important. While the decision to adopt *CMP* was not unanimous, everyone at the table agreed to support the decision and to communicate to their colleagues both the decision and how it was determined. Back in schools, the reaction from teachers was generally positive: many were excited, some were resigned, and a few were resistant (some who remain so today!).

BOARD ADOPTION

Before the vote went to the district school board in April, the K–12 Mathematics Coordinator met with each member to show them the materials and carefully explain the rationale for adoption. In early April, the recommendation was presented to the school board. We shared the broad ideas behind our recommendations, described the selection process, and summarized each program and the implementation plan including implications for professional development.

Our superintendent recognized that a dramatic shift in instructional materials would require a great deal of professional development, which was not part of our district's culture at the time. Thus he recommended the addition of five district-wide professional development days, and pledged his support for the adoption. The board was unanimous in its recommendation. Only three people from the community spoke against the materials at the presentation. Many teachers and parents testified in favor of the recommendations.

The next day, however, the local newspaper's front page headline read, "New-Fangled Mathematics Adopted by PPS." The article was mostly directed at the elementary materials (the elementary committee recommended the adoption of the *Investigations* curriculum), but certainly could have applied to middle school as well. (A year later, however, the reporter acknowledged the increases in achievement at the end of the first year of using the materials district-wide.)

In addition to the front-page article, there was an editorial that was quite critical of the materials. This prompted an outpouring of letters from teachers and parents. The writer of the editorial was affronted by the large number of letters criticizing his lack of understanding of the adoption process. The thoroughness of the work of the adoption committee and the strong consensus of the committee minimized the effects of the newspaper attacks. After these initial negative stories, little about the mathematics adoption was published in the newspaper.

Public meetings were held to inform parents about the adoption. At the middle school meeting, parents from very different backgrounds attended. Several parents who themselves were discouraged in their educational experiences were eloquent in expressing positive opinions about their children's experiences with the new materials. Other than a few comments by a handful of parents, throughout the process we heard very few complaints from parents. This can be attributed to several factors. First, we communicated often and openly with families, listening carefully to their ideas, questions, and concerns. We agreed with dissenters that there was no such thing as a perfect curriculum and that the materials selected would need to be adapted for our particular circumstances and needs. Second, we were as comprehensive as possible in planning to provide support for various constituents (teachers, students, parents, administrators, board members, and various departments such as special education, talented and gifted, and Title 1). Third, we generated long-term, systemic plans. All of this proactive planning prepared us for any opposition that was expressed about the adoption of the new curriculum materials.

LESSONS LEARNED

The process we engaged in was instructive in many ways. Thinking systemically and comprehensively and anticipating issues before they arose resulted in a relatively smooth and effective adoption process. We learned that involving teachers in a fair and representative fashion was essential. It was necessary to surface teachers' beliefs and assumptions as well as develop common understandings and shared language about research on what improves student learning, including instructional practices and materials. Engaging administrators in understanding the adoption process, the new curriculum, what to look for in the classroom, and how to support teachers was another critical factor. It was necessary to connect with and inform parents at several points in the process. Finally, having an infrastructure for supporting the adoption (including planning for ongoing professional development) was important to the success of the adoption. Additionally—

- Having many teachers involved in the adoption process created enthusiastic support for the curriculum and empowerment of teachers as they discussed important issues in teaching and learning. For some, however, the shift from a traditional textbook series to *CMP* conflicted with their beliefs about how students learn and how teachers should teach. Some of these teachers continue their opposition to *CMP* today. Thus addressing the question of how to bring non-supporters on board is one that requires continuous work.

- Early on in the adoption process, it is critical to surface teachers' beliefs about the nature of learning, teaching, students, teachers, and mathematics. Doing mathematics, reading and discussing articles and research together are effective ways to do this.
- Before beginning to examine instructional materials, it was important to read and discuss articles and research on instructional practices and content to provide a frame for the evaluation. This helped teachers be more reflective and make more considered decisions. It took time, but the time was well spent. As a result of our experiences, we would recommend up to a year and a half for this kind of careful piloting and analysis of materials in order to make a principled choice of curriculum around which teachers can rally.
- Piloting materials being considered, especially those unfamiliar to teachers, is important and challenging. It was time consuming for teachers to learn a new curriculum, particularly one that looked different from what they were used to teaching. Additionally, teachers need to collaboratively plan and debrief each new unit with their colleagues. Some of the pilot teachers in our process were able to meet to plan lessons together. Others found the logistics of doing this daunting. During our recent high school adoption process, we used what we learned to create a more effective structure to support pilot teachers.
- Parent education about reform curricula is imperative. During the adoption, we held a series of public meetings to inform parents. On reflection, an initial meeting at the beginning of the process would allow the district to share current data and standards, explaining why they are embarking on an adoption process. This would also provide an early opportunity for feedback from parents about what's working and not working with the current adoption, as well as what they hope to see in a set of instructional materials that would support them at home with their student. A mid-point check-in as well as a formal meeting towards the end of the adoption are also important. We did not do this during our adoption process, and believe it would have been valuable.
- Communicating with building administrators is essential for garnering support for and maintaining a standards-based curriculum. Since administrators leave and new ones come on board, it is important in an adoption process that administrators are represented and are regularly updated on the process and progress. At the time of textbook adoption, our district was structured in a way that promoted efficient and effective communication, which supported

the adoption process. Principals met regularly by level (elementary, middle, and high), enabling TOSAs to attend leadership meetings with principals. During these meetings they explained the adoption, involved principals in learning what to look for in the mathematics classroom and interpreting what they saw, and discussed ways to support teachers, especially those who were struggling.

- Having an infrastructure—from planning for sustained professional development over several years to TOSA help in creating documents—to support the adoption improved its chance for success. It takes a tremendous amount of work to effectively implement a new adoption, and the more this is recognized by those with fiscal authority to support the adoption, the better.
- Anticipating the nuts and bolts issues involved in a new adoption makes the transition smoother. This includes distributing and labeling books, manipulatives, and calculators, suggesting grading schemes, sharing tips for managing notebooks, and so forth. Having a plan in place to address the myriad of details that must be attended to in launching a new curriculum was enormously helpful.
- Not every teacher in every school is interested in the adoption process, but all need to be informed. Having communication structures in place to assist in the flow of information is critical.

CLOSING REMARKS

There were several key components of our middle school mathematics adoption that contributed to its success. First was a realistic timeline for completing the process. We took a full school year (from the first committee meeting in September to the recommendation in May), which was almost not long enough. In hindsight, beginning the process in the spring with the goal of making a recommendation the following spring would provide additional time over the summer for adoption activities, such as contacting publishers and getting materials for review delivered.

Including a mathematics teacher from all (instead of just some) of our middle schools, as well as including various departments and groups (Special Education, ESL, Talented and Gifted, and administrators) is a practice we recommend to others. Having a representative group at the table is important.

Formulating common understandings and consensus decision making unified the group and led to a successful outcome. We found that the process of reaching these points provided members of the committee with effective professional development as members enhanced their mathematics content and pedagogical knowledge.

Piloting materials was instrumental in ensuring that the materials were given a thorough evaluation. It is one thing to read curricula, and something quite different to actually teach using unfamiliar materials.

For a variety of reasons, we implemented the *CMP* curriculum all at once and would recommend this practice to others, rather than phasing it in over a period of years. Doing so meant that all teachers were learning something new at the same time, which contributed to the development of a network of professional learning communities. This also made providing professional development more efficient and effective. A benefit for students was that, generally speaking, only once were they placed in the situation of spending a year with a teacher who was encountering the curriculum at a particular grade level for the first time.

Finally, we developed a systemic plan for adopting, implementing, and sustaining the effort, and we made mid-course corrections as needed. We anticipated and responded to parent issues, met with the teacher union and school board members ahead of time to explain the process and hear their ideas and concerns, and kept everyone—students, teachers, administrators, various departments, and parents—informed.

Many teachers were and continue to be enthusiastic about *CMP*, noting that an increasing number of students (including those traditionally underrepresented in higher level mathematics courses) are successful on various measures of achievement and are taking more mathematics and higher levels of mathematics in high school. Teachers report that *CMP* increases their mathematical content knowledge for teaching and respects their ability to make informed pedagogical choices. Many teachers comment how they appreciate that *CMP* does not attempt to "teacher proof" its materials.

In various surveys, teachers indicated that the *CMP* curriculum:

- logically sequences worthwhile mathematical tasks
- emphasizes sense-making, generalization, and justification
- develops conceptual and symbolic knowledge
- involves students in doing mathematics
- supports students in connecting important mathematics ideas and concepts to the world outside the classroom
- incorporates technology appropriately
- promotes student thinking, reasoning, and understanding

Our longitudinal achievement data speak volumes about the success of the effort and the quality of the *CMP* materials. Since the implementation of *CMP*, test scores have risen steadily and we are closing achievement gaps, especially in 6th and 7th grade where students are typically not tracked. Notably, we did not experience an "implementation dip" during the first year of the adoption. We believe that adopting the *CMP* materials has helped us to consistently improve our mathematics instruction while raising students' achievement and closing achievement gaps.

APPENDIX: CRITERIA FOR EVALUATION OF MIDDLE SCHOOL MATHEMATICS MATERIALS

Mathematics Content

____ The materials address each strand of mathematics adequately and in depth.
____ The instructional materials address all PPS Standards and grade-level expectations.
____ The program addresses both conceptual understanding and procedural knowledge.
____ The program includes real world applications and contexts that are appropriate to a wide range of student interests and backgrounds.
____ The program focuses on larger and connected mathematical topics, rather than disconnected and isolated procedures.
____ The materials include content that is appropriate and challenging for all students.
____ The mathematics in the program becomes progressively deeper and broader through the grade levels, offering opportunities for practice and review, but avoiding unnecessary redundancy.

Evidence (reference books and page numbers where applicable):

Instructional Approaches

____ The structure of the program provides sufficient time for students to explore and investigate mathematical concepts in depth.
____ The material offers opportunities for students to build understanding through the use of manipulatives, physical and visual models, technology, discussion, practice, and self-reflection.
____ The program includes multiple representations of key mathematical concepts and multiple entry points for students to develop deeper understanding of these concepts.
____ The program includes opportunities for students to be engaged in whole group, small group, and individual tasks.
____ The program provides opportunities for students to work out and communicate their thinking.

Evidence (reference books and page numbers where applicable):

Assessment

____ The program facilitates the use of on-going multiple assessment techniques including material suitable to be used for classroom samples for the Certificate of Initial Mastery.
____ The material helps teachers assess student understanding as well as procedural knowledge.
____ The assessments are tied to instructional strategies so that the results of the assessments can help teachers plan their program.
Evidence (reference books and page numbers where applicable):

Organization of the Materials

____ The major math topics are clear and coherent and the intent of the lessons is clear.
____ The organization of the materials is clear and manageable.
____ The program includes a variety of teacher background materials that address both the math content and the instructional models used.
____ The lessons are easy to prepare, the directions clear, and sample solutions are presented.
____ The use of manipulatives, computers, and classroom supplies is appropriate and reasonable.
Evidence (Reference books and page numbers where applicable):

Evaluation Scale for Middle School Mathematics Materials

4 This program clearly implements the goal. There is strong and consistent evidence that this goal is fully developed within the program.

3 This program implements the goal most of the time. There is evidence that the goal is usually developed within the program.

2 This program recognizes the goal but fails to implement it in a consistent and thoughtful way. There is some evidence that this goal was considered within the program.

1 This program offers no clear correlation to the goal. There is insufficient evidence that this goal is represented within the program.

CHAPTER 6

THE CRITICAL WORK OF A REVIEW COMMITTEE IN CURRICULUM ADOPTION

Leslie E. J. Nielsen
Issaquah School District, Issaquah, Washington

As the brand new mathematics Teacher on Special Assignment (TOSA) in Issaquah, WA, I was thrilled on behalf of the middle school teachers to learn that the district was approved to conduct a middle school mathematics program review. Our teachers had become increasingly frustrated with the mathematics program, our middle school mathematics scores had been rather flat, and we were overdue for new curriculum materials. Facilitating the Middle School Program Review became my main responsibility for the 2005–06 school year. In my role as a TOSA, I serve as the Mathematics Curriculum Specialist. Prior to this, I had taught in the district for seven years at both the middle and high school levels. Being new to facilitating an adoption, I asked for help from my colleagues in the curriculum department and used a model they had developed as the framework for our review of middle school mathematics programs.

In Issaquah, we call this process a "program review" rather than adoption. The name in itself implies that we are looking at programs in an effort to find the very best match for our students and community. In other districts

A Decade of Middle-School Mathematics Curriculum Implementation, pages 55–66

and states where I have taught, an adoption committee examined textbooks and made decisions based on questions such as: "How does the book look? How appealing do we think students will find it? Does the table of contents match our syllabus and what we think is important? Are the teacher materials easy and friendly?" Although these questions were all valid, they do not necessarily reflect a systematic review process connected to research on how students learn mathematics. In Issaquah, we used a research-based process and thus addressed the above questions and more.

GETTING STARTED

The first step in conducting the program review was to form a committee. We held regular meetings of secondary mathematics department leaders from across the district, and that group was asked to help recruit teachers for the committee. The hope was that we would have two teachers from each middle school and one teacher from each high school participate in the program review. We came close to achieving this goal, with only one middle school lacking a participant. Teachers were compensated for time spent outside of the school day and had the opportunity to earn university credit or continuing education credit through their participation.

The committee's first meeting was an all-day retreat at an off-campus facility. Our first order of business was to adopt a Charter and Code of Collaboration. In our district, all workgroups must have a charter, drafted by the facilitator, but edited and agreed upon by the committee. The Charter (Appendix A) explicitly spelled out our responsibilities and scope of authority. We referred to this document when we ran into questions throughout the process. It also spelled out that members of the committee had a responsibility to communicate with their colleagues and principals and report that feedback to the committee. The group also framed the Code of Collaboration (Appendix B), a subset of the charter. It was particularly important to the group that we avoid "parking lot" conversations on the side that might sabotage the work of the committee. Instead, we agreed that concerns should be brought to the group as a whole and that we would work toward full consensus. We defined consensus as follows:

> *Consensus* is a decision that has been reached when *most* members of the team agree on a clear option and the *few* who oppose it think they have had a reasonable opportunity to influence that choice. *All* team members agree to support the decision.

Our work to establish what we meant by consensus became critical when we reached the point of making difficult decisions. The group was able to work through any disagreements and remain collegial. This was important because when the program review was complete, the work of implementation would be just beginning!

After agreeing on the Charter and Code of Collaboration, we were ready to dig into the research. Our goal in looking at the research was to develop a unified philosophy of what middle school mathematics experiences should be. From that we would also develop a mission statement to guide our future work. In preparation for this meeting, I had sifted through the mathematics education research and developed a short list of readings that addressed a range of topics. I structured our review of these readings using a jig-saw approach. We formed small groups, each responsible for reading one article and reporting back to the committee. To examine research on learning, we read a chapter from *How People Learn* (Donovan, Bransford, & Pellegrino, 2000). We read an article (Ellington, 2003) on the use of calculators to stimulate conversation around the use (or non-use) of calculators in middle school mathematics. To help us focus on teaching ALL children well, we read selected chapters from John Van de Walle's (2000) book, *Elementary and Middle School Mathematics, Teaching Developmentally.* Our final focus was on an article called "Supporting Students' High-Level Thinking, Reasoning, and Communication in Mathematics" (Henningsen & Stein, 2002) which led to a rich discussion of how we could maintain a high level of cognitive demand in our classrooms.

The other element of the research part of the day involved looking at our own student data. We examined our Washington State Assessment data, disaggregated by mathematical strand, ethnic group, gender, and by the date students were first enrolled in the district. We then took on the task of developing a vision of what a balanced middle school program should be. At this point something wonderful happened with the group. During the development of the vision, I stepped back from facilitating the group and the teachers took over. The big chart paper was spread out, the markers were in hand, and in-depth conversations about student learning, based on the research, ensued: "What is best for students? How can we divide it up so it makes sense? How are these things interrelated?" It was exciting to watch the committee members create a diagram showing the elements of a balanced program (see Appendix C) and then craft that into the following mission statement:

> Our middle school mathematics program is designed to support all students in meeting or exceeding Washington State Standards. To this end students are held to clearly-defined, high expectations and participate in ongoing assessment, within a positive, student-focused learning environment, with the support of appropriate instructional resources and strategies.

At our next meeting, another all day retreat, we reviewed the Washington State Content Standards. We discussed the Grade Level Expectations (GLEs) and agreed that they would be the basis of the mathematical content that we would be looking for in mathematics books. We discussed having a separate "technology" standard, but decided that technology is woven throughout the GLEs. We also discussed computation, which is often a contentious topic! We discussed how use of calculators impacts students' computational abilities, and agreed there should be balance of problem solving, number sense, and paper-and-pencil computation. We also agreed that computation should not be a gatekeeper for students doing higher-level mathematics.

Our next task was to create forms that we would use to evaluate materials. The group agreed that we needed three types of forms. The first was based on our vision of a balanced curriculum, which we used for the initial screening of materials. The second form was a standards trace in which we made sure that the GLEs were addressed in the materials. We used it for the second round of screenings. The third form was intended for classroom observations; however, one of our members, a former principal, suggested that classroom observations really do not tell that much about a program, and in the end we decided not to include them in our process. (This decision was mostly due to time and proximity. When we got down to two programs, there were no districts anywhere near us using them.) The group split into subgroups to work on one set of the forms. When possible they had examples from previous program reviews in Issaquah and from outside sources. Creating and editing the forms took the remainder of the day.

After meeting for six weeks we had still not touched a mathematics book! The members of the committee were getting a little anxious to dig in. However, during this time, I was receiving TONS of mathematics books for our review. At the start of the process, the schools in the district were closed to publishers. That meant that all communication had to go through our district central office. This is done to free teachers up from disturbances to their classrooms. It also keeps publishers from contacting teachers directly and building alliances within the district. At the district level, we put the word out that we would be doing a middle school adoption and would look at all submitted materials. We received 17 different sets of materials from a wide range of publishers, and all of those materials went through the initial screening process.

REVIEWING MATERIALS

The next three meetings involved going through materials and evaluating them using the Initial Screening Form to create a list of semi-finalists. Then the Standards Trace Form was used to further narrow down the list. Members of the group worked diligently with the forms to rate books. We all looked through the materials, read through lessons, solved mathematics problems, and read the support materials as we filled out the forms. Sometimes it was difficult to rate a book because the teacher features seemed fabulous with many resources, such as supplemental work sheets and websites and lots of teacher friendly information, but in the end we could tell that the format would revert to a teacher-centered approach. We were committed to finding a mathematics program in which students were actively involved in learning mathematics.

Eventually we were able to narrow down to two curricula that we decided to pilot. One of these was *Mathematics in Context* (National Center for Research in Mathematical Sciences Education & Freudenthal Institute, 1997–1998), which was the curriculum that was subsequently chosen. Conversations leading to this selection were intense and required a great deal of listening and thinking. Throughout this process members of the group showed a great respect for each other, and were extremely thoughtful in their interactions. Nevertheless, it became problematic in this part of the process that we had an under-represented middle school. The representative from that school was unable to come to a couple of meetings and was surprised by the direction the committee took in his absence. This occurred even though I had communicated with the teachers and principal at that school to keep them informed of the committee's progress. One affirming bit of news that I shared with the committee after they decided on the two finalists was that the two programs they selected were the two top-rated programs by the AAAS Middle School Textbook Analysis (AAAS, 2000).

The next step for the committee was to coordinate the piloting of the materials. The group decided that each program should have a three-week window for the pilot. Teachers at each school spent time with the materials and chose the unit that they felt was most appropriate to replace the content that they would have been teaching in the existing program. More teachers participated in the pilot than on the program review committee. These teachers all participated in a one-day training for each of the two programs. The trainings were conducted by consultants for the programs and arranged by the publishers. At these trainings, teachers were introduced to general features of the program and given specific instructional suggestions for the units that they were going to teach. It was a little tricky planning the sessions, since many different units were being discussed. During the pilot, teachers agreed to keep a double entry journal formatted as follows:

Date	What you did in class today.	Reflections on the day.
	Here you will record what you did with your students. (short bullets to describe the context)	Please record your reflections on today's lesson with special regard to: • How did the lesson go from the teacher perspective? • How did the lesson go from the student perspective? • What evidence did you have during the lesson of student learning?

At the conclusion of the pilot for each program, teachers were asked to reflect on their experience and send their reflections to me. They were also asked to respond to the following questions:

1. What mathematics did your students learn during this unit? What evidence do you have that learning occurred?
2. To what extent did these materials challenge and engage students at their ability levels? Were struggling learners engaged? Were students challenged?
3. Did the materials provide opportunities for formative assessment as well as summative assessment? Did it provide you ways to identify ideas and misconceptions that your students had?
4. To what extent were students involved in "hands-on" activities? Were the activities relevant and engaging?
5. How was technology used in the unit you piloted? Was it a genuine use of technology that enhanced the program rather than being an add-on?
6. Were students required to explain their thinking and support their answers?
7. What did you specifically like about this program?
8. What did you dislike about this program?
9. What resources would you need to teach using this program?
10. What training do you feel you would need to teach using this program?

After each pilot, we met with all teachers to debrief the experience and made a list of pros and cons of each program based on the teachers' reflections. These data were one piece of evidence we used to make our final decision.

In our district, parents are active partners. Although they did not serve on the Program Review Committee, we did create opportunities for them to give us feedback and information on their priorities. The committee drafted a letter that was sent home via US mail to the parents of each student in classes that were piloting, and parent information sessions were scheduled at each middle school, either at a PTSA meeting or at the Site Council meeting. I went to these meetings accompanied by the Assistant Superintendent for Teaching and Learning. At each of these meetings, we asked parents the question: "In a Standards-based Mathematics program, how can we best assist you in helping your child?" At each of these meetings, we explained what is meant by standards-based mathematics, and how the program review was unfolding. We showed them the two programs and we discussed the resources for parents to help their children within each program. Parents are eager to help their students and become frustrated when they do not understand the mathematics or can not see the mathematics underlying the activities in the books. Feedback from parents was shared with the review committee.

At the conclusion of each pilot, we also asked students for feedback. We had a questionnaire for each student and their feedback was informative. I went through all of the questionnaires and made a list of the most frequent responses. In hindsight, I would ask fewer questions next time and have a numbered scale for some, but the original responses were fun to read. In response to the question, "What did you like about the *Mathematics in Context* book?" students told us that they enjoyed the contexts used in the book: "I like how the researchers took 1500 pairs of sons and fathers. It seems like they have made a lot of effort into this book." Students also liked the approach in *Mathematics in Context*: "It was simple and straight forward." "I liked that it explained the concepts before you practiced them." "I liked that we learned it in different ways besides paper and pencils (example, gumdrops and toothpicks)." "EVERYTHING! I like the fact that it was fun & easy @ the same time." There were a few comments that compared to the previous book that had been field tested: "It gave more clear ideas than the last, and gave me a better idea on what to do," and it was "harder," "explained better" and "gave more clear ideas than the last one." The comment that many teachers on the committee really enjoyed was "Most of the problems were easy, but still required thought. Very interesting. I wanted to read ahead." None of the teachers on the committee had ever had a student say they wanted to read ahead in a mathematics book before.

REACHING CONSENSUS

In April, armed with teacher responses, student opinions, parent requests, and all of our own opinions and beliefs, we started working towards consensus. The next meeting was very difficult. We knew we were working with two exemplary programs, and each had strengths that were evident. At the end of the day it became clear that we were not ready to reach consensus. We agreed that all committee members would go back to their buildings and get more feedback and that we would meet again as soon as possible. In the two-week window between this meeting and the next, I made trips to all of the middle schools to meet with teachers and make sure everyone had seen all of the materials. I encouraged teachers to let their representatives on the committee know what they thought.

The committee met again at the end of April, and we reached consensus at the end of the day. Being a group of mathematics teachers, we developed a mathematical model to help us make our decision. I divided the information we had into five categories: two represented student and teacher feedback and three pertained to our initial vision of a balanced program (clear and high expectations with ongoing assessment; spiraling approach that emphasizes the inter-relationships between mathematical domains; and materials and resources). We divided into two small groups with equal representation from across the district. Each group had 5 points that they could award for each element. For each element, the group could give all of the points to one of the programs or the other, or split between the programs. For each point awarded to a program, the group had to fill in a sticky note that described the reason why the program received that point. The groups were asked to make detailed notes to explain their thoughts to the rest of the committee. At the end of this activity, *Mathematics in Context* [*MiC*] had a slight advantage, and it became clear that most teachers were ready to go with *MiC*. The group listened carefully to the concerns of teachers who were not in favor. Before we made our final decision, I asked the members to take a break and walk outside for a few minutes to reflect on the decision. I absolutely did not want to have anyone regret the choice once it was made.

Once we made the decision to go with *Mathematics in Context*, we had to prepare for the next step, which was presenting our recommendation to the Instructional Materials Selection Committee (IMSC). This is a group of parents, teachers, community members, and administrators from across the district that reviews all adoption recommendations prior to recommendation to the School Board. The Middle School Mathematics Program Review Committee presented *Mathematics in Context* to the IMSC and returned several times to address questions. This process was more lengthy than usual since the publisher was in the process of coming out with a second edition. We were working from proof copies of the new edition and that caused some concern for IMSC. On June 29, the IMSC approved the adoption of *Mathematics in Context*, and on July 13 it was approved by the School Board as a consent agenda item. This was a very quiet ending to a long, exciting process.

The last work for the review committee to address was to make recommendations for implementation. The committee recommended that all three years of the new program be implemented simultaneously. They strongly urged that all teachers participate in a three-day workshop in August before the school year started and be provided on-going professional development during the year to occur on five release days per grade level. But that's another story...

REFLECTION—LOOKING BACK

Now that it has been a little more than a year since we completed the adoption process, I can look back from the perspective of one year of implementation. Overall, our implementation has been quite smooth. Administrators at the building and central level in the district have commented to me on how well it has gone, which has been welcome feedback. As this was my first time facilitating the review process, the perspective of those with more experience was very helpful.

That being said, there have been molehills, which have felt very mountainous. One of these occurred as the result of one building not being adequately represented on the committee. One school had only one representative, and for various reasons, he missed two meetings. In the course of those two meetings, we narrowed our list of choices from about eight down to two finalists, and the program that he preferred did not make the list. As a result, he has not taken ownership of this implementation, and teachers at his school are not implementing as energetically or enthusiastically as one would have hoped.

Another molehill has been that not all members of the committee were able to stick with the Code of Collaboration. When we reached the difficult stage of making the final decision, some people had to give in. We reviewed the definition of consensus, and I made every effort to make sure everyone had been heard, and that everyone could live with the decision. And yet, I hear discussion and rumors in the buildings about how the decision was made that show a real belief that somehow I had an agenda and pushed it. The building that wanted the other program has been unhappy this year, and that has gotten in the way of the implementation going well for all of them. And yet, they are implementing. Those teachers who have had an open mind are having a good year, and have warmed up to the implementation of *MiC*.

One of the things we did in the process that ensured success was the parent communication piece. Most parents in the district felt informed throughout the process and were not surprised by the implementation. They were pleased to have seen the materials before the decision was made and glad that we asked for their input on what they wanted in a program, even though they did not get to participate in the committee's decision making process directly. When parents have expressed an interest in having a more direct role, we have reminded them of their representation on the IMSC.

Overall, our review and adoption process was very thorough. In conversations with parents, teachers, and community members this last year who have called wanting information (or perhaps to complain), I was pleased to be able to tell them that the decision to adopt *Mathematics in Context* was not a random one. It is wonderful to be able to tell them that a team of our very best teachers completed a systematic and careful study, and that we can say with confidence that the committee's recommendation is the very best middle school mathematics program for the students of Issaquah.

APPENDIX A
Charter: MS Mathematics Adoption Workgroup

Authorized by	District Improvement Agenda and Teaching and Learning Services
Purpose	To review and produce a comprehensive implementation plan for Middle School Mathematics to include grades 6 through 8 and a scaffolding plan for students in grade 9 who are not ready for the first year of the existing high school mathematics program.
Time Frame Goals	• Review of literature on the best practices in mathematics education. • Examine our data to see how it can inform us. • Review of Washington State GLEs with the NCTM Standards and align to Issaquah Standards. • Create tools to evaluate materials. • Review and evaluate materials, select two semi-finalists. • Field test materials. • Evaluate data from field testing. • Select materials. • Plan implementation of materials.
Membership	Up to 17 teachers. Three from each middle school, one from each high school and the mathematics TOSA.
Selection Criteria	Willingness to serve and attend meetings regularly. Ability to communicate effectively with building teachers and principal and report feedback to the committee.
Facilitation	Leslie Nielsen, TOSA with Teaching and Learning Services Role: Prepare agenda, monitor meeting time, guide discussions, find research materials, acquire adoption materials
Approach	Each team member should be prepared to: – Effectively bring issues to table – Work toward having interests understood – Learn from other representatives – Work from a student perspective – Take learning back to building teachers and principal.
Scope of Authority	This group will advise and make decisions as appropriate. The decisions of the group will be taken to the Assistant Superintendent for comment and approval.
Support	The District will provide compensation for participants at the staff development per hour rate.
Report	Minutes will be posted after each meeting on the district intranet

APPENDIX B

Code of Collaboration/Facilitator Norms

- Maintain sense of humor and have fun.
- Listen to and show respect for the views of others.
- Attend fully—engage in listening.
- Bring problems with options whenever possible.
- Speak one at a time; minimize disruptions.
- Share responsibility for the team's success.
- Give everyone a chance to participate.
- Be honest! Don't just say what you want others to hear.
- Converse outside of this workgroup with the intent of bringing something constructive back to the group. No sabotage. If you have concerns, say it here or don't say it at all.
- Use consensus for decision process as agreed upon.

APPENDIX C: Elements of a Balanced Mathematics Program

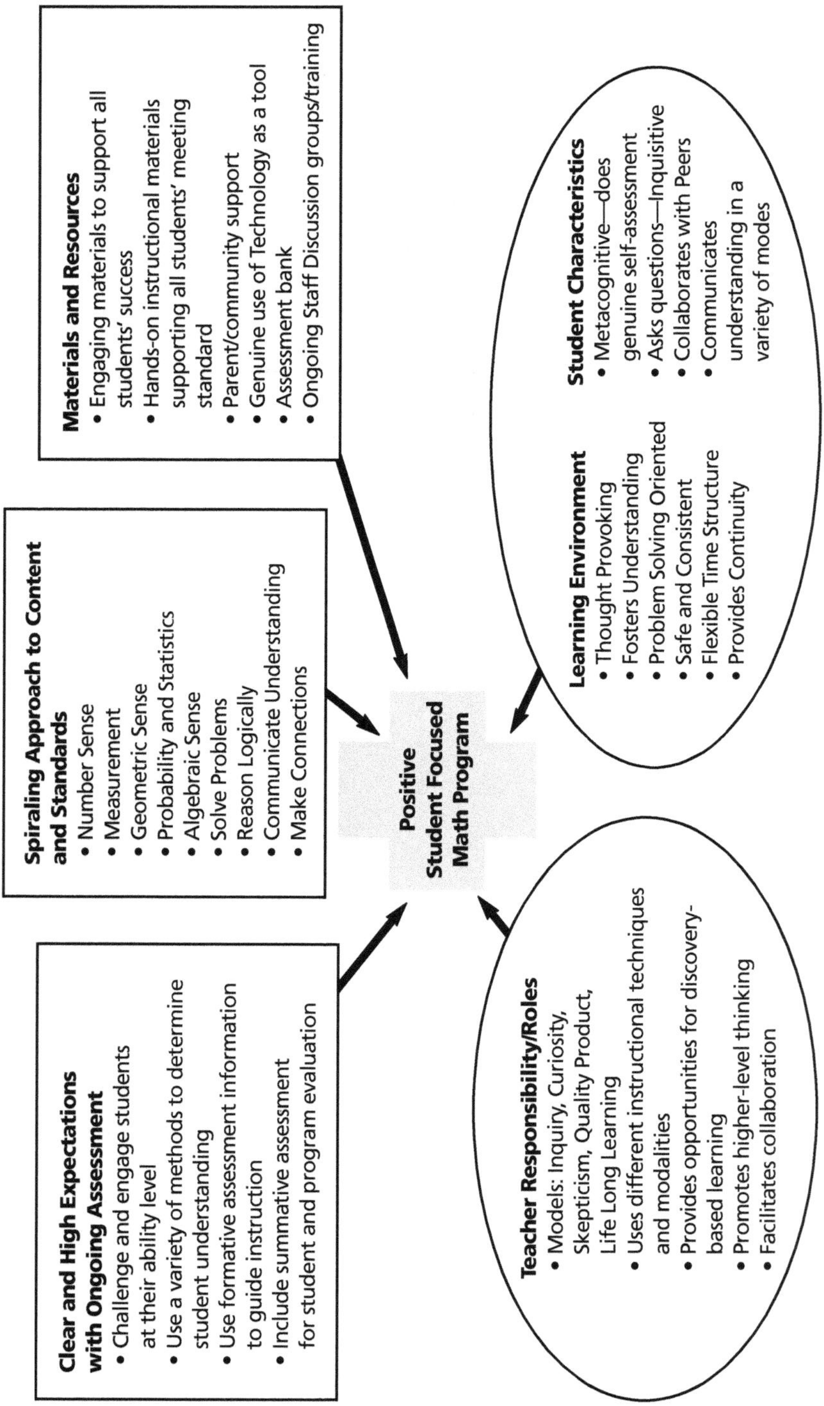

CHAPTER 7

INFORMING A STATE ADOPTION PROCESS THROUGH CURRICULUM IMPLEMENTATION

Murrel Brewer Hoover
WVSTEM Center@Marshall University, Huntington, West Virginia

This is the story of a unique adoption model based on a state-level professional development initiative using the *MathScape* curriculum materials (Education Development Center, 1998). The Mathematics Education Reform Initiative for Teachers (Project MERIT) was a 5-year project funded by the National Science Foundation[1] and supported by the West Virginia Department of Education to improve the teaching and learning of middle school mathematics and to enable teachers to make informed decisions about the selection of curriculum materials to support instruction. To achieve this goal, the project sought to develop networks of teacher-led learning communities to sustain and support teachers as they changed their instructional practice to reflect recommendations of the NCTM *Principles and Standards for School Mathematics* (2000). The project selected *MathScape* as a curriculum to use as the focus for professional development for all grades 6–8

A Decade of Middle-School Mathematics Curriculum Implementation, pages 67–77

mathematics teachers in West Virginia because it is a standards-based curriculum and closely aligned with the state framework. The expectation was that having used the *MathScape* curriculum as a model, teachers would be more likely to adopt one of the four NSF-funded programs during the next textbook adoption.

PROJECT PARTICIPANTS

As with most long-term programs, a variety of participants were involved in Project MERIT. There were three types of key players: MERIT leaders, full-time mentors, and project teachers. Although MERIT leaders included mathematics and mathematics education faculty from institutions of higher education in West Virginia, most of the leaders were drawn from the ranks of teachers who were involved in the project. At least one teacher leader from each of the 55 county school systems and more in higher population counties were involved in the project. MERIT leaders were trained to implement *MathScape* and then provide professional development to their colleagues in the use of the curriculum. They also served as role models within their schools for the use of *MathScape* and as a resource and source of support for colleagues participating in the project.

The project also provided for four full-time mentors, each responsible for working with one-fourth of the mathematics teachers in the state. They supported the work of the MERIT leaders and provided critical connections and colleagueship throughout the project. The mentors served as coaches for the mathematics teachers, spending time in classrooms observing, modeling and team-teaching various *MathScape* lessons. Over 600 project teachers from across the state participated in MERIT. These teachers were organized into two separate cohort groups. The first cohort, consisting of approximately 225 teachers, was kept smaller than the second to give the project leadership experience in designing and delivering professional development centered on the implementation of *MathScape* and related efforts to support that implementation. This first cohort was initially very enthusiastic about participating in the project. Many of these teachers became strong advocates for standards-based instruction and promptly implemented the *MathScape* materials in their classrooms. The second cohort of approximately 450 teachers began in the third year of the project. Although this cohort was more reluctant than the first group (some were skeptical about standards-based mathematics programs), many of these participants ultimately implemented the materials as well. They were sometimes referred to as the "coerced but converted."

CURRICULUM IMPLEMENTATION AS A VEHICLE FOR CHANGE

Initially, the aims of the project were focused on the MERIT leaders. They were involved in on-going professional development activities throughout the course of the project. Early professional development sessions were focused on the content and pedagogy of standards-based mathematics in general and the *MathScape* curriculum in particular. Later sessions included techniques for coaching and mentoring peers and planning professional development. The philosophy of the project called for teacher leaders to use the *MathScape* materials with at least one class of students and to draw on these experiences in providing professional development to other teachers. Thus teachers would lead professional development sessions only after they had experienced the use of the materials and made changes in their own instructional practice as recommended in the *Standards* (NCTM, 2000). For some MERIT leaders there was concern about leading professional development sessions, and many who thought they would never do so, turned out to be the best facilitators.

Developing MERIT Leaders

A key component of the professional development for MERIT leaders was their participation in summer institutes and follow-up sessions aimed at implementing the *MathScape* curriculum materials. During summer institutes they worked through *MathScape* units as learners themselves, completing investigations and discussing the mathematical ideas with each other. In addition to this immersion in the material, the teachers also studied various content strands to see how individual units related to the overall development of particular content across grades 6–8.

During the first summer institutes and follow-up sessions, teachers received professional development to implement three of the seven *MathScape* curriculum units. They were expected to use the three *MathScape* curriculum units as a supplement to their previously-adopted instructional materials. They began the school year with a unit, met regionally in learning communities to discuss successes and challenges and to analyze student work. They then taught the next unit, met regionally again, and then taught the third unit. Units were identified so teachers could alternate between a unit with which they were familiar and a new unit. During the second year of implementation teachers received a waiver to replace their county adopted textbook with *MathScape.* Most leaders reported the first year as the most difficult—not only because the material was new, but also because it was very difficult to move between the use of curriculum materials with two completely different philosophies.

By providing high quality standards-based instruction using *MathScape* materials, teachers found they could engage all students at different levels using tasks requiring higher-level thinking skills. Unlike, in the past, when teachers would attend a professional development session and use selected activities in their classroom, *MathScape* gave teachers the resources they needed to present a coherent, comprehensive program. The materials made changing to standards-based instruction much easier as teachers could return to the classroom and immediately use the knowledge gained in professional development. The philosophy of using curriculum implementation as a vehicle for change is embodied in the adage, "Appeal to the heart and the head, but give the hands something to do."

Additionally, the MERIT leaders reported an increase in and deepening of their mathematics content knowledge as a result of the professional development and implementation of the *MathScape* materials. Teachers studied their grade level units in depth, but the units were also addressed in the larger context of studying a content strand. This was particularly important for many of our 6th-grade leaders. For example, Sandy, a 6th-grade MERIT leader with multi-subject K–8 certification teaching in a middle school was assigned to teach 8th-grade mathematics. Her story below illustrates the strong content embodied in the curriculum materials played a role in helping her obtain advanced certification in mathematics through Algebra I and gave her confidence in her ability to teach the mathematics.

> I was scared to death to come to the first meeting (MERIT). Not only an entire room filled with math teachers but an entire hotel, all those math teachers... and me... I now feel that I have a great math family... I've learned a great deal of mathematics from teaching *MathScape* and MERIT has made it possible for me to receive my certification, which I will get at the end of this semester. I am in the process of finishing the MERIT calculus class. I have made great friends and I have learned how to teach math.

Learning Communities—a Vehicle for Sustaining the Change

Another important component of the project was the establishment of learning communities to support and sustain changes being made in the classroom. MERIT leaders attended monthly learning community meetings facilitated by various leaders within regions across the state. Members of the learning community shared their successes and challenges in implementing the *MathScape* materials and used student artifacts as the basis for assessing classroom changes. Within these learning communities, teachers enacted action plans consisting of a student focused learning goal (such as increasing student communication); the benchmarks or steps for reaching that goal (often based on the 2000 NCTM *Standards*); the actions they agreed to take in the classroom; and the documentation (often in the form of student work) they agreed to bring back to the next meeting. Because the *MathScape* units were designed to include numerous opportunities for students to communicate their thinking through problem solving, writing or discussion, the curriculum materials supported the work of the teachers in enacting these action plans. During year 2 of the project, MERIT leaders established learning communities among teachers within a local area or school. These learning communities became, and remain, the vehicle for sustaining change. During the last year of the grant, learning communities became a statewide initiative from the Department of Education, thus giving credibility to the work of the project.

Overcoming Challenges—Re-Thinking Past Practices

Because we teach the way we were taught, it is a difficult change for teachers to shift their instructional practice from being a presenter of knowledge to a facilitator of learning. But this is exactly the kind on change MERIT leaders were experiencing as they incorporated the *MathScape* materials into their instruction. The power of these curriculum materials to support teacher change was illustrated by one of the MERIT leaders during a presentation at the state mathematics meeting. Lillian openly talked about coming to the first MERIT meeting, sitting with her arms crossed, a sour look on her face, and saying, "This will never work—my kids can't do it." She reported that by having the on-going support of a MERIT mentor, she stayed with the professional development and that now she will never go back to teaching the way she taught before the project. During her talk, she held up a *MathScape* book and said, "This is what made the difference for me... and... if I'm 53 years old and I can change, you can do it too!"

Addressing the needs of all learners was another challenge faced by MERIT leaders. The state was moving toward a more inclusive model for exceptional needs students, so from the beginning, the project advocated for special educators to attend professional development with their regular education mathematics teachers. This story about Lynn and Lee Ann who began teaching in a collaborative classroom illustrates the success the MERIT leaders enjoyed as a result of using *MathScape* materials that were accessible and engaging. Lynn wrote:

> Our resource room teacher, (another MERIT teacher), brought her class into my class as we began the *Making Mathematical Arguments* unit. I thought we would just see how things went. So on Tuesday, an additional 10 students, 1 teacher, and 1 aide joined my mathematics class. This has been one of the best weeks ever. You would have thought that these students had been working together for weeks. After the first day, I wondered if things would continue—they continued all week. All of the teachers joined in as the groups worked. You could not pick out the special education students. They began to participate more each day. It was a win, win, win situation. It is amazing what can be accomplished when teachers work together. This should be a 'no brainer.' Students benefit by working together, why shouldn't the same be said for teachers?

There are several aspects of the *MathScape* curriculum that fostered this kind of collaboration and learning. The activities were accessible to all students; everyone could find a way to get started. The investigations invited student participation and discussion in multiple ways—speaking, building, drawing, writing—so all students could contribute their thinking to the discussions. Furthermore, although some students could take the ideas to more advanced levels than others, everyone gained something mathematically from the experience.

Sustaining Implementation—Expanding Professional Development Opportunities and Gaining Support

As MERIT leaders moved into the second year of implementation with *MathScape*, it became clear that they were ready for another level of professional development that went beyond gaining familiarity with the curriculum materials. Questions about assessment, use of collaborative groups, and ways to invite and support student thinking in the classroom became more pressing as teachers began to see the potential impact of the *MathScape* materials and instructional approach. As a result, outside consultants with particular areas of professional development expertise began to play an important role. As one example, during the second year of implementation, a facilitator with expertise in collaborative learning in mathematics conducted a session with the MERIT leaders. In addition to providing techniques to successfully implement collaborative groups, the facilitator helped the teachers recognize that all students can learn mathematics. Many MERIT leaders felt this professional development opportunity was a turning point in their implementation. The strategies they implemented enhanced the effectiveness of the instructional materials. These strategies then became an integral component of the professional development they later delivered to cohorts of MERIT teachers.

Although a few sessions in the first two years of the project were designed to prompt teachers to think about their role as MERIT leaders, it was not until near the end of year two that the project seriously focused on teachers leading professional development for other teachers and recruiting others into the project. In subsequent years, spring and fall conferences were held for MERIT leaders to focus specifically on developing leadership skills and providing professional development.

In the final three years of the project, regional planning meetings were held each year as a strategy for encouraging the MERIT leaders to advocate locally and in order to gain county buy-in to support teachers through professional development. MERIT leaders coordinated teams from their county consisting of county superintendents, Title II coordinators, professional development coordinators and others at the county and/or school level who made decisions about professional development offerings in the county. At the first of these meetings, the MERIT leaders were somewhat reluctant and reserved about planning with county personnel; however, as the meetings progressed and opportunities were presented for them to share the results of classroom implementation, they became strong advocates for standards-based mathematics and professional development for teachers in their counties.

Providing Teacher Professional Development

During year three of the Project, the MERIT leaders began to lead professional development sessions for other teachers in their regions. Lessons learned from their own experiences caused us to alter the professional development plan. For example, with MERIT leaders, three or four units were introduced during a weeklong seminar with only one day to develop each unit. The teachers found that it was overwhelming to focus on multiple units and that they needed to spend major time reviewing the materials just prior to teaching each unit and because of the length of time between the professional development and the time they actually taught the unit.

The planning for the first summer institute for teachers took these issues into consideration and the new four-day institute allowed the participants to experience one *MathScape* unit in depth—including cooperative learning strategies, teaching reading in mathematics, writing, and assessment as they were embedded in the unit. All these strategies were built into the first unit as it was developed including explicit conversations about the impact of these strategies on implementation. With this background, the participants then returned to their classrooms ready to implement the first unit. Subsequent unit development occurred as one-day seminars spread throughout the school year just prior to when teachers were expected to teach the unit. One day was sufficient for planning to teach a unit once teachers fully understood the changes needed to implement the *MathScape* materials and had the opportunity to try it in their own classrooms. As with the leaders, the expectation was to use only three units the first year and implement all seven units during year two.

The two cohorts of teachers committed to 18 days of professional development—nine each year for two years. In addition to the summer institutes and one-day seminars for professional development on unit implementation, teachers met for the equivalent of three additional "support" days. These were scheduled on alternating months with the one-day unit planning seminars in order to provide some "just in time" support for the unit they were currently teaching. The project maintained an online Web board through which teachers across the state discussed implementation and other issues. Additionally, this provided a location for resources to be shared and accessed electronically. Over the course of the project, MERIT leaders provided professional development to more than 950 mathematics teachers (grades 6-8) throughout the state of West Virginia.

TEXTBOOK ADOPTION PROCESS AND OUTCOMES

The project had been organized to end during the instructional materials adoption year for mathematics. However, a strategy was put into place to provide some training to assist districts in making their mathematics program selections. In years 3 and 4 of the grant, Project MERIT partnered with the K–12 Mathematics Curriculum Center at EDC to facilitate "Considering New Curricula" seminars. From this training, their extensive knowledge base and background experience using the *MathScape* curriculum, their understanding of the need for change and recognizing what that change looked like, and their placement in the role of leader and advocate, the MERIT leaders were poised and ready to take their place on state and county textbook adoption teams. As a textbook adoption state, a statewide team was assembled to write the criteria by which instructional materials would be reviewed and recommended for the West Virginia textbook adoption list, which counties use to determine their selection. MERIT teachers, leaders, and administrators who were among the members of this committee used their expertise to advocate for and design separate criteria for three categories (Standards-based Components, Technology & Manipulatives, and Mathematics Content) to be used in the evaluation of all material. Previously in mathematics as well as in other content areas, instructional materials needed only to meet the criteria regarding content. This shift toward using multiple categories was a major step and sent a message to publishers that we were looking for instructional materials to support the implementation of the NCTM *Standards* (2000). Instructional materials were first reviewed for their components related to the NCTM *Standards.* Only those that met 80% of the criteria in that category moved on to be evaluated in the next category and the process continued in the same manner through each of the three categories.

Another statewide committee was formed to actually review instructional materials according to these categories for inclusion on the state list of materials for adoption. Once again, it proved beneficial that members of this committee included MERIT teachers who were experienced with the mathematics content and pedagogy of the *MathScape* curriculum. On several occasions MERIT teachers drew upon their experiences to advocate for the inclusion of an NSF-funded curricula. In doing so, they drew on their expertise to point out how the materials met a particular criterion when it was not obvious to those without experience using such materials. Their experience at the middle level also enabled them to make connections with elementary and high school level NSF-funded curricula.

Once the state adoption list was constructed, MERIT leaders provided professional development to nearly all of the 55 county adoption teams. The professional development focused on making standards (NCTM, 2000) relevant, understanding the importance of instructional materials and how NSF-funded curricula were different from more traditional materials. Participants viewed videos and "did the mathematics" to understand the significant differences between a traditional approach and a standards-based approach in teaching for understanding.

Early in the adoption process, publishers recognized the emphasis that counties were placing on NSF-funded curricula. One publisher began to offer a classroom set of NSF-funded curriculum materials if counties would adopt the traditional materials or a classroom set of traditional with the adoption of the NSF-funded curricula. In some instances the offer appeased adoption teams who needed a "middle ground" solution. Unfortunately middle ground actually undermined the efforts toward standards-based materials since, given the option, most teachers chose to stay with past practices. Some counties, which were reluctant to commit to the requisite professional development for a standards-based program, passively encouraged the more traditional option. However, a few counties denied the offer of traditional materials and adopted one of the NSF-funded curricula exclusively.

The outcome of adoptions statewide was mixed. All 55 counties currently have *access* to one of the NSF-funded curricula, but are actually using these materials to various degrees. In the majority of cases where teachers received professional development through MERIT and continued to be supported in changing their practice, NSF-funded curriculum materials are being fully implemented. This is not the case, however, for teachers who did not participate in the Project or who have not received subsequent support

TABLE 7.1 Number of Counties Adopting Different Curriculum Materials

NSF-Funded Curricula	Publisher-Developed Curricula
Connected Mathematics—9 counties	Prentice Hall—5 counties
MathThematics—3 counties	McDougal-Littell—4 counties
MathScape—11 counties	Glencoe—6 counties
Mathematics in Context—1 county	Holt—13 counties
Waiver—1	Unreported—2

for implementing standards-based instruction. Table 7.1 summarizes the textbook adoption outcomes at the middle school level.

CLOSING REMARKS

Two important requirements for providing meaningful, effective professional development are time and money—Project MERIT provided both. While middle school mathematics was the emphasis, time and money were available to support teacher leaders and for them to support their colleagues. As a result of this effort, districts were more informed about standards-based curriculum and instruction as they reviewed and adopted curriculum materials.

There were many lessons learned throughout the duration of Project MERIT, particularly about the many stakeholders in the processes of adopting and implementing curriculum. For example, when the project ended and time and money for professional development were no longer available, districts responded differently. In schools and counties with administrative support and strong teacher leaders—those who learned to not only change their teaching practices, but also learned to advocate for high quality mathematics professional development and instructional materials––the programs have continued to grow. In other locations without such leadership, traditional teaching practices and curriculum materials have prevailed. Without strong advocacy and leadership, even in the face of improved test scores, teachers who are isolated will succumb to the approach and pressure to "cover" the "basics" instead of taking the time to develop content and understanding or be tempted to once again teach from a collection of activities instead of a coherent standards-based program.

Other results indicate the positive impact Project MERIT and the use of the *MathScape* curriculum has had on continued efforts to reform mathematics in West Virginia. Anecdotal success stories related to MERIT and the implementation of the *MathScape* curriculum materials also abound. The 7th graders in one classroom, after finishing a unit called *Buyer Beware* (Bousquet, Janssen, & Brodesky, 2005) and learning about unit pricing, "took on" the Frito Lay company because they thought they were being charged too much for the snacks in the machines. Kenna, a parent of a 6th grader, whose family was out to dinner at one of those restaurants where the tables are covered with paper and kids are given crayons to draw pictures, looked over and instead of drawing, her 6th grader was creating patterns with numbers and patterns with shapes and writing the algebraic expression to go with them.

Several positive indicators point to continued efforts to reform mathematics in West Virginia. An emphasis at the elementary level on NSF-funded curricula in half the counties in the state will certainly provide support for implementation of these curricula at the middle level. Several districts are demonstrating their understanding of the important role that administrators play in the change process and have instituted professional development programs for administrators in the district. A statewide emphasis on 21st century learning skills continues to keep the focus on standards-based instruction and the pressure to implement NSF mathematics curricula at the forefront. And, finally, it is encouraging to recognize MERIT leaders who are now in leadership positions such as principals, mathematics coordinators and numeracy coaches; state-level leaders in assessment and Mathematics/Science Partnership grants; mathematics specialists at university professional development centers; and numerous MERIT leaders in key positions on county and state-level committees who have the opportunity to influence decision-making regarding mathematics and professional development at all levels.

The work of Project MERIT has been intense yet fulfilling work. Stories about student engagement in mathematics from teachers and parents make a compelling case for continued implementation of NSF-funded curricula. Even the "coerced but converted" teachers have come to recognize the increased understanding, retention, and problem-solving abilities of their students. Most project teachers indicate that they will never go back to teaching the way they taught before and will continue to advocate for the implementation of instructional materials like the NSF-funded curricula at all grade levels.

"Never doubt that a small group of thoughtful,
committed people can change the world.
Indeed, it's the only thing that ever has."

—Margaret Mead

NOTE

1. Project MERIT was funded by the National Science Foundation under teacher enhancement-professional development (ESI-9911928).

CHAPTER 8

CURRICULUM IMPLEMENTATION THROUGH TEACHER COLLABORATION

Chip Sharp
Columbia Public Schools, Columbia, Missouri

The year Columbia Public Schools (CPS) began full implementation of *Connected Mathematics* (Lappan, Fey, Fitzgerald, Friel, & Phillips, 2002), I became the full-time mathematics coordinator for grades 6–12. I entered teaching as a second career and after eight years as a classroom teacher, I became the mathematics coordinator on a part-time basis. As a former junior high and high school mathematics teacher I brought a variety of experiences to the position of secondary mathematics coordinator. These experiences included using the NSF-funded secondary mathematics program *Core-Plus* (Coxford et al., 1998), serving as president of our state mathematics association, and collaborating as a writer of the Missouri Curriculum Framework and related documents.

This is a story about the implementation of *Connected Mathematics* (*CMP*) in Columbia Public Schools. In this chapter I describe our mathematics program pre-*CMP*, our early efforts to implement *CMP*, and implementation activities that occurred through our partnership with the Show-Me Project.

A Decade of Middle-School Mathematics Curriculum Implementation, pages 79–86

I end the chapter with recommendations for mathematics coordinators so that others might benefit from our experiences.

BEGINNING TO IMPLEMENT *CMP*

Prior to our implementation of *CMP*, we used a more traditional text. Many teachers, however, were displeased with the mathematics program. Some felt that they were speeding through the material and only "covering" it rather than teaching concepts in depth. Furthermore, it was common practice for teachers to pick and choose the topics that were covered, a structure that did not allow teachers in subsequent grades to build on students' common experiences or understandings. This situation did not promote a consistent, quality mathematics program for the majority of our middle grades students—our top students did well, but many other students were lost in the transitions from grade level to grade level.

When *CMP* became available commercially, CPS teachers began using one or two units as replacements to textbook chapters with similar content. Although many teachers struggled to implement *CMP* in the way intended by the authors, they were even more frustrated with the district-adopted textbook. When time for a new textbook adoption arrived, teachers overwhelmingly voiced a desire to adopt and implement *CMP* because they had experienced its potential for student learning through their use of the replacement units.

My own classroom experience with implementing *Core-Plus* at the high school level gave me insight to the challenges the middle grades teachers would encounter as they implemented *CMP*. At the time I did not feel that most teachers completely understood the changes that would be required, but I knew that without materials like *CMP*, it would be even more difficult to move towards student-centered instruction in mathematics. Even with their desire to use *CMP*, teachers encountered difficulties with the implementation, largely because they did not fully understand the program's intent or structure. Many remained focused on "doing" a unit and expected that students would reach "mastery" by the end of the unit. We realized that for successful implementation, teachers needed to develop new ways of thinking about learning and teaching mathematics.

Like many districts, our use of *CMP* began on a shoestring budget. Although we offered after-school meetings for professional development, the level and scale of this support were not enough to provide all teachers the help they needed. The support for implementation of *CMP* primarily consisted of publisher presentations, short overviews of the materials, and after-school inservice sessions to address issues about using student-centered learning materials. However, teachers really wanted sessions specifically aimed at "how to do this." Thus attendance at professional development sessions was minimal and the implementation was not moving forward district-wide. I felt we were floundering and was frustrated with the lack of support from the district and the lack of incentives for teachers to come together and work on instruction. I felt that teachers wanted to learn more, but needed more support to encourage their investment of time. Little did I know that opportunity was just around the corner.

PARTNERSHIP WITH THE SHOW-ME PROJECT

Two years into the implementation of *CMP* we were asked to join the Show-Me Project as one of their partner schools. This partnership allowed us to create greater opportunity for teachers to become immersed in the learning necessary to fully implement the program. In the four years we were involved with the Show-Me Project, we provided two types of professional development opportunities for teachers: weeklong summer institutes and *CMP* Saturdays.

The weeklong summer institutes were designed to give teachers an intense immersion into the *CMP* program. Although local teacher leaders provided support for this professional development, I also felt that it was important for our work to be informed by someone with expertise in using *CMP*. Our connection with the Show-Me Project enabled us to secure a top quality facilitator who had worked closely with the curriculum developers. The structure of the summer institutes was two-fold: mornings were spent in large-group discussions directed at "big picture" issues and work in the afternoon focused on what the teachers would be teaching the following year. Teachers worked in grade-level groups to become familiar with the curriculum materials at their grade level. These sessions were designed to follow up the morning experiences so that the entire day was connected.

Fortunately we were able to use the same *CMP* facilitator for 2 years. This gave our teacher leaders time to grow in their understanding of the program and become confident in leading discussions on their own. We also involved two teachers from other districts using *CMP*, and our thinking was challenged and stretched by these teachers who were not restricted to CPS past history. They offered new insights and ideas that were sometimes outside the comfort zone of our teachers, but the summer institute was a safe environment and teachers became more willing to think differently about teaching and learning mathematics.

CMP Saturdays were designed to target the specific needs of teachers. For the first two years, sessions focused on what to expect when using these curriculum materials. Teacher leaders walked teachers through lessons and indicated the pitfalls and challenges they might encounter when teaching a unit. Often teachers worked through some of the investigations, following the same model as the summer institutes. *CMP* Saturdays were particularly helpful to teachers new to the district and most teachers were content to continue attending sessions of this type. But after two years, I felt we needed to shift gears. From my observations, teachers were not sufficiently digging into the mathematics of the units. One clue to this was our lack of consistent use of the Summary portion of the lesson. I felt that this was the result of teachers not being comfortable with the important mathematics with which students should make connections. Thus, for the last two years the focus of *CMP* Saturdays was on understanding the big ideas of middle school mathematics. We came to a consensus about the big ideas at each grade level and thought about what that means in student friendly language. We examined student work samples as a way to become more consistent in our expectations of students' performance. As a result of these efforts, teachers now talk about the mathematics that students understand and use examples of student work to provide evidence of this understanding. Many teachers have raised their expectations of what students can achieve and are frequently blown away by the level of student understanding that results when they (the teachers) get out of students' way. Because of our work with the Show-Me Project, teachers view themselves as learners who still have much to learn. To some, this is discomforting but it is also very empowering and has created an energetic tension to improve instruction.

Our partnership with the Show-Me Project enabled us to offer stipends and in-service credit for teachers who participated in professional development. Accordingly, these sessions became more attractive to teachers and during the four years of the partnership, over half of the middle grades teachers regularly attended professional development for *CMP*. This was a great improvement over the first two years of implementation. Teachers have grown to appreciate the support provided by *CMP* Saturdays and continue to attend these sessions even though the Show-Me Project has ended. Recognizing the need for ongoing professional development, principals have provided support for a new initiative that allocates release time during the school day for teachers to meet. Attendance at these school day sessions is mandatory, ensuring that all teachers have access to some professional development in mathematics.

TEACHERS COLLABORATING TO LEARN

There are many stories that I could tell about the implementation process. One story of particular importance pertains to how teachers have come together to work as a team to improve their own knowledge during the process of learning how to teach with the *CMP* materials. Prior to our implementation of *CMP*, it was not a part of the culture for CPS teachers in different buildings to work with each other or seek advice for improvement. In part, the teachers did not know each other very well, but more importantly, the teachers really did not have any motivating factors that would cause them to seek out other teachers from across the district. Occasionally, teachers would meet at in-service workshops, but their conversations typically addressed surface-level aspects of teaching mathematics and tended to focus on sharing favorite activities. However, with the implementation of *CMP*, a need arose for teachers to really talk with each other about instruction and learning. The opportunity to work with the Show-Me Project came at the perfect time because teachers were ready to examine their instructional practice. I believe this is why a good number of teachers volunteered to participate in the professional development that was offered. With a mathematics program like *CMP*, students often demand that teachers help them understand the mathematics. Our teachers found that students did not sit passively waiting for the teacher to tell them something. Teachers were suddenly confronted with this "in-your-face" need of the students, which led to an increased need for strategies that would develop students' understanding. Students were asking questions that teachers found they could not answer. Thus, a need to learn motivated many teachers to meet with other teachers across the district who were teaching the same grade level.

CMP Saturday sessions became a forum for teacher collaboration. Along with examining pedagogical issues, teachers worked together to understand the embedded nature of mathematics in *CMP* units. For example, order of operations is not a stated objective in *CMP* until the 8th grade. However, there are many opportunities to explore this convention in lessons at the 6th- and 7th-grade levels. We found that many teachers were not aware of the depth or variety of mathematical ideas within the materials. Also, teachers did not take advantage of opportunities to make mathematical connections within units and across the grade levels. During professional development sessions, teachers worked together to increase their understanding of mathematics in general as well as the mathematics addressed in *CMP* at their grade level.

Another important collaboration that occurred involves special education teachers and their use of *CMP*. In CPS, some special education students attend "regular" mathematics classes taught by a mathematics teacher, but have a special education teacher in the class to support their learning. This is known as a "Class within a Class" setting. When we first began using *CMP*, most special education teachers felt that their students could not do this type of mathematical work; they believed it was too hard and required too much reading and writing for their students to be successful. However, a few special education teachers attended *CMP* professional development sessions and began to see the real potential of *CMP* for their students. They created a different culture of expectation and found that their students could succeed in the "Class within a Class" structure. In addition, these special education teachers implemented *CMP* in their self-contained mathematics classes, in place of the low-level remedial materials they had been using. They also expected these students to take the same common assessments that other students were taking. This shift, initiated by a few special education teachers who attended *CMP* professional development, has promoted positive results both in the classroom and on "high stakes" assessments.

CHANGES I HAVE OBSERVED

Over the course of implementing *CMP* in our district, I have seen teachers grow from needing to be the source of knowledge to where they want to structure instruction so that students are benefiting from the learning. I see more teachers getting out of the students' way and allowing them to engage with the activities so the students have a chance to really understand the mathematics. Six years ago when I walked down the hallways I saw students sitting in rows at various states of engagement, watching their teachers work at the blackboard. Now when I visit middle schools I see teachers asking better (less leading) questions, questions that do not diminish the challenge of the problem. As I observe classrooms, I see teachers allowing students to discuss the activity and conclusions without feeling the need to share what they (the teachers) know as the answer. When I see groups of students discussing the mathematics and not waiting on the teacher for help, it is clear that these students have been given the opportunity to learn in a much richer manner. While we still have a ways to go with student achievement, we are encouraged by the ways in which our middle grades students are coming to understanding mathematics.

I see more teachers valuing the need to learn more. I see teachers understanding and enjoying the tension of needing to be a continuous learner. I also see teachers valuing students being involved in their own learning. As the district coordinator, I am convinced that without these teachers working together over the past four years, I would not be seeing classrooms as I described above. I continue to think about the changes I have observed in the CPS teachers as a result of their hard work; I see teachers rethinking what it means to teach. The excitement teachers show as their students demonstrate greater understanding than expected is so enjoyable for me.

SUGGESTIONS FOR DISTRICT COORDINATORS

Implementing NSF-funded mathematics curriculum materials requires dedication, time, some money, and a lot of patience. As I reflect on the implementation of *CMP* in our district, I have a few pieces of advice for anyone who is coordinating such an implementation from the district level. I believe this advice would apply to any implementation effort, no matter which NSF-funded mathematics curriculum your district chooses.

- Get teacher buy-in! Even if it is from just a few teachers to begin. Because the implementation is going to drive the need for improving teaching and student learning, it needs to have advocates at the teacher level. I found that a curriculum coordinator may not be perceived as “a teacher,” so having other strong teachers involved really helps. Establish an advisory group of teachers to facilitate communication about how things are going in the classrooms, to share problems that are being reported in their schools, and to brainstorm possible solutions to those problems.
- Work to make connections with universities to gain assistance in thinking about and developing learning opportunities. These connections are also good ways to tap into other funding sources that may be necessary.
- Be mindful of teachers’ needs in providing professional development and recognize that their needs will change throughout the implementation process (see Chapter 19, this volume). This will engender greater teacher buy-in and participation in professional development.
- Encourage leadership development. By allowing teachers to take on more of a leadership role, the entire group becomes empowered with better thinking and possibilities for improvement. These issues are too large for one person to know all, be all, and do all. Acknowledging this and encouraging greater teacher participation in leadership roles creates more ownership in the implementation process.

Although I have not discussed administrators or parents in this chapter, it has been my experience that these two groups can greatly impact a mathematics coordinator’s work. I offer these additional suggestions as necessary considerations when implementing NSF-funded programs.

- Get administrative buy-in! Principals and counselors must recognize the need for change and understand how mathematics instruction is different from in the past. They will often be the first people that concerned parents (or the media) will contact.

- Communicate with parents. This is a long-term need, as you will continue to have parents who are new to the curriculum. We are now conducting more Parent Nights than we did at the beginning of implementation, and in two different versions. One version focuses on how parents can help their child at home and the other focuses on issues with the program itself. While some of these meetings can be difficult at times, it is rewarding to see the value that teachers and parents place on our efforts. I have found that parents may not always agree with our efforts, but I believe if we maintain the focus on improving student understanding, then students will be the winners. I also suggest that districts consider having a Parent Advisory Group to elicit parental feedback and promote trust in the program.

CONCLUSION

In my role as district mathematics coordinator, I watched the middle school teachers in Columbia Public Schools grow in their knowledge of mathematics and in their ability to work with students to increase mathematical understanding and achievement. Change is neither quick nor easy, a thought that often conflicts with the American culture of expecting quick, easy solutions. Historically, we have not supported middle grades students in learning mathematics on a deep, connected level; nor have we supported teachers in adapting their instructional practices so that students can learn in this way. I hope this chapter can inform the work of other district coordinators who are trying to support teachers and increase students' mathematical understandings at the middle school level.

CHAPTER 9

CURRICULUM IMPLEMENTATION

Teachers on a Journey

Carol Den Otter
University of Sioux Falls, Sioux Falls, South Dakota

Implementing standards-based curricula is a bit like leaving your homeland to come to a new country. As I reflect back on our district's experiences with *Mathematics in Context* [*MiC*] (National Center for Research in Mathematical Sciences Education & Freudenthal Institute, 1997–1998) it occurs to me that an analogy can be made to my grandparents' immigration to America. They knew something about where they were headed, but certainly had no idea about how long and rough the sea voyage would be when they got on the ship. Nevertheless, they were willing to take a risk because they believed that they could not better themselves by staying in their home country.

Just as my grandparents left their homeland in search of a better life, so too did our middle school mathematics curriculum committee seek to leave traditional mathematics textbooks behind in search of a better way of teaching mathematics. As the committee read research, studied documents from national and international mathematics assessments, discussed the National Council of Teachers of Mathematics (NCTM) *Principles and Stan-*

A Decade of Middle-School Mathematics Curriculum Implementation, pages 87–96

dards for School Mathematics (NCTM, 2000), and reflected on past practices, it quickly became apparent to many committee members that traditional mathematics just was not in our future.

While my grandparents believed that it was necessary to leave their homeland to give themselves a chance at a better life, their siblings chose to stay close to home where they were comfortable. Some committee members harbored those same feelings as they looked through sample sets of materials. The familiar two-page spreads, enrichment worksheets, and practice sheets coupled with some fancy manipulatives and computer software lured them into arguing for staying with something traditional and comfortable. Their hope was that a new look and more practice sheets in addition to standardized test practice would certainly help their struggling students. Meanwhile, the "explorers" on the committee continued to examine the depth of mathematics found in the *MiC* materials. Even though the lesson pages were not as attractive as other textbooks and the *MiC* sales person did not offer free pens, meals, or a polished sales presentation, the majority of the committee members began to consider the possibility of adopting *MiC*.

The adoption committee, made up of middle school and high school teachers, administrators, and a parent, met for several days in the summer to review sample texts from every NSF-funded curricula. As a result of that meeting, the committee decided to have a small group of teachers pilot one unit from three different middle grades curricula in the fall and report their findings back to the committee in December. At that critical meeting, after hearing lengthy reports from teachers regarding their students' experiences with the pilot, the committee agreed unanimously to adopt the *MiC* curriculum. The school board officially approved the adoption in January. Little did they or the teachers know about the land that lay ahead. They were about to discover new and sometimes rugged territories.

SUPPORTING TEACHER USE OF MiC

Professional development for teachers started as soon as *MiC* was approved. In late spring, *MiC* consultants came to the district and led a weekend workshop attended by all the middle school mathematics teachers. Teachers were paid to attend the meeting. Some of the veteran teachers voiced concerns about the adoption of new materials. Since they had been successful teaching out of traditional books, they would have preferred to have a more familiar approach. The teachers who had served on the adoption committee were eager to get started and many taught *MiC* units before the implementation officially began. The explorers were getting a feel for the land, liked what they found in this new curriculum, and stumbled on a treasure trove of student questions and insights about mathematics that emerged in their classrooms. Students demonstrated a level of engagement with mathematics that was refreshing, which led to new questions about the role of problem solving, communication, and reasoning in middle grades mathematics.

Besides working with the teachers, *MiC* staff helped educate the middle school principals. An *MiC* consultant met for a half day with administrators to introduce them to the program and to answer their questions. How were basic skills and homework addressed in *MiC*? Is the reading level suitable for English-language learners and special education students? Does *MiC* challenge and prepare gifted students for higher mathematics? What is a typical *MiC* lesson and what should we expect to see when observing a classroom? They were grateful for the training they were given and were better equipped to support the efforts of classroom teachers. As administrators observed lessons during the first year of implementation, their support for the program continued to grow based on positive reports from teachers and parents about the renewed student excitement in doing mathematics. Because they trusted the master teachers in their buildings who supported the curriculum, they were solidly behind the adoption and implementation.

During the summer before the implementation, a group of teachers and I attended a week-long *MiC* summer institute in St. Cloud, Minnesota. As the curriculum coordinator for our district, I recall being bothered by the fact that local teachers were paid a daily stipend to attend the institute and our teachers received no stipend. It seemed we were asking a lot from our poorly paid teachers to give up a week of their summer without compensation. Nevertheless, the teachers who attended were excited about learning how to teach reform mathematics and appreciated the opportunity to network with other mathematics teachers. The lack of stipend was no big deal to them. They came away more determined than ever to make this program work.

This institute offered teachers powerful learning experiences during the summer months when they did not have to worry about time away from their students. As teachers worked through the curriculum with their colleagues and *MiC* consultants, they became more confident in their abilities to successfully implement the program. While we were there, one of the trainers offered two comments that I often used during our implementation. The first was "be kind to yourselves during the first year." And the second was "If you find that it is taking more than six weeks to complete a unit, at some point during a unit you are going to need to just stop. There are many exciting and interesting activities you can do, but there is only so much time." The explorers left St. Cloud with new friends, new mathematical knowledge, new pedagogy, new colleagues, and a renewed vision for the future of mathematics education.

During the first year of the implementation, all teachers attended a half-day of training every ten weeks that was facilitated by an *MiC* consultant. During these sessions, the teachers worked through the problems in the next unit they were going to teach. Each grade level met at a different time; for instance, the sixth grade met during a morning, the seventh grade met during an afternoon, and the eighth grade met the following morning. The time of day was switched during the next round so teachers did not always miss the same classes. Numerous teachers were grateful for these opportunities to learn more about the curriculum, to share what was happening in their rooms, and to learn from their peers. There were also a few teachers who were clearly frustrated and would have jumped ship if given the opportunity. Homework assignments and tests were common concerns. Trying to complete a lesson in our district's 40-minute class period was nearly impossible. These issues were resolved primarily through discussions that emerged in these bi-monthly meetings. Teacher concerns could have been eased much sooner if we had planned upfront to offer all teachers time to meet together and collaborate in unit planning throughout the school year.

For teachers who had help and support from their principals and colleagues along the way, the journey became much easier. In some schools, teachers would meet by grade level each week to plan their lessons, to share what was working, and to resolve problems. Because common planning time was scheduled for the teachers in one of the schools, this joint planning time was easy to accomplish. For others, they had to work to make it happen. This meant that one day a week, they either had to come early or stay late in order to find a time that worked for them. Others chose to go at it alone. Of the five middle schools, the most progress toward fully implementing *MiC* was made at the buildings where common planning was provided or where teachers were willing to work outside the school day to plan. The opportunity to work in a learning community paid off. In retrospect, I wish we had encouraged teachers to visit each other's classrooms to observe and discuss lessons and student responses. This missing piece–to learn from each other's practices–would also have strengthened the implementation of *MiC*.

During the first year, there were many complaints from parents because they did not know how to help their children do the homework. A few professional people such as doctors, engineers, and lawyers told teachers and the central office that this curriculum was horrible. They said if they could not figure it out (with their advanced degrees), how could we expect the students to do the work? They were afraid their children would fall behind and lose out on opportunities to attend prestigious colleges and universities. The local tutoring business said that the *MiC* curriculum was not teaching students what they needed to be learning. Yet, in spite of this negative response from some parents, the teacher leaders never wavered in their belief in the program and I did my best as the coordinator to keep everything moving in a positive direction. There were at least three factors that helped to keep me focused: interaction with the *MiC* curriculum developers, my 20-plus years as a mathematics educator, and the master mathematics teachers out in our schools.

To respond to the concerns being raised by some parents, we encouraged schools to have parent open houses to help the parents better understand the program and appreciate what students were doing in mathematics classrooms. These open houses were successful in achieving greater parent buy-in at most schools. There were also times during parent-teacher conferences in which teachers were verbally attacked, but they quickly learned how to skillfully handle these situations. Those teachers who firmly believed in what they were doing to improve student understanding and excitement of mathematics had fewer upset parents. When these parents questioned the decision to use *MiC*, those teachers were well prepared to defend the curriculum, because they believed in it and could see the results in student work and the mathematical insights students shared on a daily basis in their classrooms.

During the final quarter of that first year, parental complaints had escalated in one of the middle schools. One of these parents happened to be a school board member. Looking back, if we had taken the time and effort to involve the school board members more during the adoption process, we may have avoided this confrontation. As a means to dispel rumors and alleviate fears about the effectiveness of this approach to teaching and learning mathematics, I met on six different evenings with parents and tried to explain why we needed to change the way mathematics is taught and what we hoped to accomplish. It is worth noting that the meetings were attended by a small but vocal group of parents. Parents were encouraged to ask questions and after hearing more about the need for change and the curriculum, most of the parents left the meetings feeling more satisfied with the implementation of *MiC*.

FACING THE CHALLENGES

Change is always difficult and it often takes time and lots of communication to help everyone feel comfortable. We implemented the *MiC* curriculum at all three grade levels—6th, 7th, and 8th—the first year. This was especially difficult for the eighth-grade teachers and their students as some of the prerequisite skills had to be taught before they could begin a unit. For students who had sat in a traditional mathematics class for seven years, it was hard for them to change from having direct instruction to one that relied on inquiry. Students who had been successful with the traditional approach wanted to continue on that path and would often come to school and tell the class that their parents did not understand this new mathematics, so why should they? Some students appreciated the routine of doing exactly what was modeled for them by the teacher, and then if they did not understand, they could rely on their parents to explain the procedure to them later at home. However, some students who had formally struggled trying to understand the teacher's efficient explanation were now finding success when mathematics finally began to make sense to them and was connected to the real world.

Meanwhile, the teachers were trying to come to terms with what homework to give the students. This was different than anything they had done in the past. *MiC* consultants did the best they could by providing answers to any questions the teachers asked. The *MiC* support team received several phone calls and email messages asking for help during those early years. As the curriculum coordinator I needed their guidance on more than one occasion and they were always willing to offer support and encouragement.

While the middle school teachers were working hard to make the implementation a success, high school teachers were already hearing rumors from parents who were worried that their middle school students would not be prepared for high school mathematics. During the fall semester of the second year, the high school teachers were up in arms. The incoming ninth graders did not know the distributive property as well as they had in the past. Two of the three department chairs from the high schools demanded a meeting with the eighth-grade middle school teachers. During that meeting, they were not open to listening to any of the remarks coming from the middle school teachers. Except for the "explorers" in their ranks, they believed our curriculum was a disaster even though some of the respected high school faculty did advocate for the *MiC* curriculum. One department chair tried to win his colleagues over by telling them that he appreciated the new knowledge students had. For instance, he did agree that they may be weaker in one or two areas, but he could teach them those skills. He was thrilled to see them more willing to work at problem solving and to try to figure more out on their own. What a gift he was!

We gave the high schools copies of the *MiC* teacher and student books and encouraged those who raised concerns to work through a few of the exercises to help them better understand what was happening in our middle school classrooms. Middle school teachers also presented overviews of various lessons during our combined meetings to help orient the high school teachers to this reform mathematics, but too many were stuck in their comfort zones and did not want to venture forth on a new expedition. In their busy lives, they had little time to sit down and review the units. It would have been wonderful if we could have had them work through some of the eighth-grade *MiC* units. Throughout the implementation, we struggled to win acceptance from most of the high school mathematics teachers and principals. I believe that our task was daunting and to this day admire the teachers who steadfastly attended those tension-filled meetings at the high school.

PROFESSIONAL DEVELOPMENT IN YEARS 2 AND 3

Each summer thereafter, a group of teachers would attend a *MiC* conference either in district or out-of-state, or both. They were always appreciative of the training they received and were excited to go back to their classrooms. Unfortunately, a few of the staff never attended any training after the first year of implementation. Perhaps there should have been training during school release time in the second and third years of implementation to offer these teachers opportunities to learn more about *MiC*. Ideally, high school algebra teachers also should have been attending the same training as the eighth-grade teachers. Unfortunately, staff development money in our district was limited and so a prime opportunity for teacher learning and collaboration was lost. Eighth-grade teachers did continue to meet with high school mathematics teachers to discuss the *MiC* program and what was being worked on in middle school mathematics. If I were to implement a reform program again, I would make sure funding was available for high school algebra teachers to attend at least the eighth-grade training sessions so they would better understand the mathematics their incoming students were learning.

Perhaps the best professional development experience I offered teachers was a week-long staff development course in the summer that focused on the cross-grade development of mathematical strands in *MiC*. The purpose of the class was to help teachers learn more about the program from a mathematical point of view. Teachers met in small groups and worked through the problems in each strand from each of the three grades. Some completed several strands and told me they learned more mathematics than they had during their college years. About one-fourth of the mathematics teachers took advantage of this opportunity. Another success story was when the district sent several middle school and high school mathematics teachers to the annual NCTM conference. By getting to know one another outside of school parameters, the teachers became more accepting of each other and this helped to relieve the tension between the two levels.

During years two and three of the implementation, a few of the teachers decided they could not become "explorers" and chose to retire early, move to the high school, or change departments. As new teachers were hired, every attempt was made to encourage them to attend a summer training institute. For those who chose to attend, their first year teaching the curriculum was easier. Each year two master teachers spent one full day training new mathematics teachers. In some cases, new teachers got to plan with a veteran teacher who helped them along the way. Again, this approach seemed to work best in schools that had weekly planning sessions for mathematics.

Every year teachers and administrators worked with parents new to the curriculum to help them understand the curriculum and to show them how they could best help their children with homework. Another crucial piece of the parent issue was the homework dilemma. As teachers learned how to deal with homework, some of the parents' frustrations with the new curriculum were resolved. Teachers continued to explain the curriculum to parents at back-to-school nights and during conferences and gradually parents began to feel more comfortable with the new way of doing mathematics. I believe that as the teachers became more confident in using the curriculum the parent issues decreased. When the explorers saw the benefits of using the new curriculum, they became believers and never turned back. In fact, the teachers even agreed to participate in the field testing of the second edition of *MiC* in spite of the challenges in becoming familiar with another set of new and revised instructional materials.

In year four, I left the district to accept a faculty position at a local university. Another member of the curriculum department was hired to do my job and the new superintendent asked the department to begin a K–12 mathematics study. Even after my departure I was pleased to find that the explorers were not willing to return to their home country. It was exciting for me to hear that the middle school teachers believed in reform mathematics and had no intention of returning to traditional mathematics even though the state mathematics test for the middle school grades was written for a traditional text. They argued, "Why change back? Our students are doing well!"

REFLECTIONS ON THE IMPLEMENTATION PROCESS

As curriculum coordinator it was my job to organize the curriculum studies and plan for the implementation of newly adopted curricula. I led the middle school mathematics study when we adopted *MiC* and this committee worked diligently from the beginning of the study to find the best program for our students. During the implementation I planned the staff development, served as the liaison between the teachers and the *MiC* consultants, supported the teachers in whatever way I could, and worked to develop better relations with the high school mathematics teachers and administration.

In reflecting on this process, I know we made the right decision, but we probably made the implementation process more difficult than it should have been. In retrospect, neither the superintendent nor the school board had much understanding of what they had approved. Their limited understanding of the nature of these materials came back to haunt the curriculum department on numerous occasions. If I were to lead another adoption and implementation, I would take the time to supply the superintendent and the board with more information about the commitment that would be required from administrators, teachers, and parents to support a successful implementation. Although this may lengthen and complicate the adoption process, this information would have offered key administrators a more authentic portrait of the potential challenges they would encounter. Furthermore, I think the committee and the curriculum director could have provided greater communication with the teachers and listened more to their concerns. However, change is never easy and sometimes it is necessary to "just begin the journey."

The success of the sustained implementation was remarkable considering that mathematics reform in our district was not a systemic effort. I say that because at the time we were using *MiC* in the middle grades, both the elementary and high school teachers continued to maintain their use of traditional materials. Students entered the middle grades with limited experiences in problem solving, reasoning, and articulating their mathematical thinking. Students transitioning into the middle grades were then given three years to learn a new way of doing and thinking about mathematics, and then they were expected to go back to a more traditional approach once they entered high school. As new principals came onboard in subsequent years, they were a bit reluctant to support teacher use of the materials. One reason for their reluctance was that they came from the high schools that did not support our decision. If I had it to do over again, I would have given the new principals the same half-day training that the others had the opportunity to attend. In fact, perhaps it would have been beneficial to bring all the middle school principals in for training as new leaders were hired so the experienced principals could share how they had been supporting teachers in implementing a new curriculum.

If I had it to do over again, I would beg my colleagues in the elementary schools to move more quickly to a curriculum that would support what we were doing in the middle schools. A solid K–8 implementation in reform mathematics would give students a strong conceptual framework on which to build their mathematical knowledge. They would be expected to think and reason everyday. They would be encouraged to make sense of their mathematics and to be problem solvers. In addition, I would fight for more staff development. We should have continued the quarterly half-day training sessions for at least the first three years. Generally, it takes three to five years to help teachers become comfortable with the changes required in classroom practice promoted by reform curricula.

As I look back on our story, I am proud of the explorers who saw a need to change, worked hard to make it happen in their classrooms, and encouraged others to buy into the program. We could not have made it happen without the training and support given to us by the teacher-consultants, developers, and professional development opportunities along the way.

CHAPTER 10

THE ROLE OF TEACHER LEADERS IN CURRICULUM IMPLEMENTATION

Emily R. Fagan
Educational Development Center, Newton, Massachusetts

Rhonda K. Weinstein
Public Schools of Brookline, Brookline, Massachusetts

Relating the story of one district's implementation can help to illuminate the many aspects of curricular change. While some elements of any district's implementation will be unique to the district, many of the experiences will be replicated time and time again in other implementations in other districts. Examining the story of a district's experiences over time can reveal the changing needs of teachers and the steps one district has taken to sustain a strong implementation. An implementation story also illustrates how a district and its teachers continue to develop through different phases of the implementation. That information will hopefully prove useful to other districts that are set to embark upon a curricular change to *MathScape* (Education Development Center, 1998), or to a similar standards-based middle grades mathematics program.

This is the story of one district's implementation. It reflects the experiences of four employees in the district: the curriculum coordinator, a veteran

A Decade of Middle-School Mathematics Curriculum Implementation, pages 97–105

sixth-grade teacher, a 7th/8th-grade mathematics teacher with seven years of experience who joined the district in the first year of implementation, and a mathematics/science teacher who joined the district in the fourth year of implementation. The three teachers worked at different schools within the district and all taught in inclusion classrooms, which were supported by special educators who taught in some of the mathematics sections. The district adopted *MathScape* in 1996 and has been using these curriculum materials for more than ten years.

THE SELECTION AND ADOPTION PROCESS

Prior to the adoption of *MathScape*, curriculum materials used in grades K–8 were fairly supportive of standards-based reform recommendations. Although teachers embraced the philosophies and approaches of these programs, there were elements lacking in the materials. Some teachers, parents, and students were dissatisfied with the materials. A shortcoming from one teacher's standpoint was the lack of information provided in the teacher guides. They did not offer enough support with respect to how the lesson should be taught. At the time, many teachers did not really know how to teach mathematics except in the traditional way they had been taught; they felt they had to figure this out for themselves. It is also important to note that the district provided a limited amount of in-service training for these programs.

When the district began to explore the adoption of a new mathematics curriculum in the early- to mid-1990s, the administrators did not have a formal process in place for materials review and selection as currently exists in the district. However, there was a strong interest in change among teachers, and there was administrative and community support for the change. The system-wide mathematics coordinators concluded that the programs being used needed to be reconsidered, and they initiated a review of other materials with the intention of selecting the best program. The upper-grades curriculum coordinator then invited teachers to join a committee to establish selection guidelines and then to review and select materials.

In the 1994–1995 school year, the coordinators formed two focus groups, and the search began for programs that would be "the best for students and the most teacher-friendly." One of the focus groups addressed curriculum options for kindergarten through grade 5; the other addressed options for the middle grades. The committees examined many of the NSF-funded standards-based programs as well as more traditional textbooks. Both focus groups were comprised of diverse groups of teachers. The curriculum coordinator overseeing the middle school group took care to include a teacher from each of the eight K–8 schools, as well as a representative from the high school and someone who taught 5th grade. In addition, she insisted on representation from the Special Education, English Language Learners, and Gifted and Talented programs.

In reflecting on the process, one member of the focus group recalled that the fit between *MathScape* and the district framework and culture seemed better than that of the other curricular options. The curriculum materials were compatible with the district's mission *to educate each student to become a responsible adult capable of contributing to the quality of life in a free and changing society. We aim to develop capable, confident lifelong learners who contribute to their community, participate thoughtfully in democracy, and achieve success in a diverse and evolving global society.* Moreover, the district's core values of high academic achievement for all, excellence in teaching, collaborative relationships, and respect for human differences aligned well with the principles and philosophy of *MathScape.* So, the committee recommended the selection of *MathScape*; the curriculum coordinator concurred, and the school committee approved the selection.

INITIAL IMPLEMENTATION AND PROFESSIONAL DEVELOPMENT: YEAR 1

The staff in this district was very savvy about the demands a new program would place on the district. In fact, at the outset of the review process, the curriculum coordinator realized that a successful implementation of any NSF-funded program would require sustained professional development. The coordinator viewed professional development as an opportunity to provide workshops in both mathematics content and student-centered pedagogy. Some other initiatives were already underway to improve mathematics teaching, and the implementation of these new materials would serve as a vehicle for deeper examination of content and instructional practice. In recognition of the work associated with this curricular change, the district developed a plan to phase in *MathScape* by implementing two or three units at each grade level each year for three years. During this transition period, the district continued to use units from the previous curriculum materials until all of the *MathScape* units were phased in.

The district developed a budget that provided ample funds for teachers to attend regional workshops. The plan began with a summer institute prior to the first year of implementation. During this weeklong summer institute, a subset of teachers volunteered to participate in unit workshops designed to prepare them to teach two or three units of *MathScape.* Then, the district provided three full release days so that all middle school teachers could attend workshops on three units for their grade level. Finally, a subset of teachers again attended another weeklong workshop during the following summer in order to prepare themselves to teach more of the remaining units. The district supported three full release days for each of the three years of the program implementation. The plan for professional development represented a significant investment in the district's teachers, the program, and the future success of the implementation.

KEEPING THE COMMUNITY INFORMED

The district's implementation plan also included family communication as an objective and outlined steps to achieve that goal. During the latter stages of the review process and prior to implementation, the curriculum coordinator, along with some administrators and teachers, held meetings in each of the schools to inform parents about the upcoming changes and to field questions from parents. These meetings were an important element in the transition to *MathScape*.

In reflecting on the reaction from the community, the curriculum coordinator recalled that parents wanted to know what was happening, and some questioned why the district needed to change curriculum materials. They were particularly concerned about how calculators would be used in the classroom. They worried that the use of calculators in the middle grades might result in weaker computational skills. In addition, because the curriculum materials looked different from what parents themselves had used, it was difficult for them to evaluate the mathematics contained within the program. These questions were reasonable and predictable, and the district worked to provide ongoing information to the community and to invite feedback from parents. Because the approach was a proactive and collaborative one, families were receptive to the information and to the change. They did not feel left out of the process. In the end, as one teacher described, "Parents were mostly happy, with the exception of a very vocal minority who wanted a traditional program." The district attempted to field concerns in a thoughtful and systematic way and was careful not to place teachers in the position of having to defend the curricular choice. The goal was for teachers to concentrate on integrating the use of the materials into their practice and for administrators, who were knowledgeable about the selection process and relevant research in mathematics teaching and learning, to address questions about the program. To this end, while teachers were welcome to discuss the program with parents, they were also invited to refer any concerns to the curriculum coordinator.

Other elements of communication to parents included mathematics breakfasts and open houses. During the first year of implementation, each teacher invited parents to a breakfast in the classroom to see some of their children's work and to engage in mathematics activities from the program. The curriculum coordinator dedicated meeting time to work with teachers at each grade level to plan for the event and to support their selection of activities. Similarly, the curriculum coordinator supported teachers in conducting a Back to School Night by allocating meeting time to plan for this event. Teachers collaborated to identify appropriate activities to share, and to review or create handouts describing the mathematics program and district objectives.

In the ten years since this implementation process was initiated, the district has instituted an even more inclusive curriculum review and selection process. Although parents had been part of previous program reviews, 2006 was the first time in at least 30 years in which parents were part of the materials selection process. Ongoing and clear communication with families is critical to any successful implementation. Educating families early on about the rationale for change helps to dispel potential concerns or opposition. Moreover, this communication is likely to generate strong supporters of the program.

SUSTAINING THE IMPLEMENTATION AFTER YEAR ONE

The district realized that teachers would benefit from ongoing support beyond the initial examination of the curricular units associated with the first summer institute. Although teachers' primary concerns at the outset of implementation pertained to the new content and pedagogy of the *MathScape* materials, over time these needs changed. Teachers' interests and concerns turned toward assessment practices, the alignment of *MathScape* to the state test, and other ways of communicating about *MathScape* to colleagues, administrators, and parents. To address articulation issues, the district formed optional cross-grade study groups, which met after school. These cross-grade groups helped teachers find support for their own practice and learn more about how the mathematics they were teaching at their grade level was further developed in subsequent grades. In addition, monthly grade-level department meetings were dedicated to collaborative planning. Finally, in response to the needs of teachers to adapt and extend materials to meet a range of learners, the district formed working groups and paid stipends for summer work on program development. All of these efforts helped to keep teachers connected around a common goal and supported their efforts to realize change in the teaching and learning of mathematics in this district.

From these implementation efforts, there emerged a group of teacher leaders who could provide ongoing support to other teachers, thus increasing the district's capacity to provide professional development and enabling the district to cost-effectively sustain the implementation. Many of these teachers were part of the subset of teachers who participated in the early summer institute workshops. Rather than hiring outside consultants, the district capitalized on resident teacher leaders' expertise, collegial relationships, and firsthand knowledge of the district culture. These teachers were able to provide workshops to other teachers, and they served as mentors to teachers new to the district. In addition, teacher leaders were able to facilitate meetings to evaluate how the curriculum was supporting district and state learning goals, and to assess the professional development needs during various stages of the implementation.

ADDRESSING STAFFING CHANGES

After the initial 3-year cycle of professional development and the full phase-in of *MathScape*, the school district needed a professional development plan for new hires and teachers moving to different grade levels. Using both the expertise of teacher leaders and outside resources, the district created a plan to provide multiple opportunities for professional development to new teachers and veteran teachers who were teaching at different grade levels.

Summer Workshops

The district invited all new hires to attend a 3-day summer workshop offered locally for area districts using *MathScape.* The district paid for the workshop but did not pay stipends to the teachers for their participation. Most teachers welcomed the opportunity for this kind of professional development prior to the start of the school year. The three-day workshop focused on the first two units in the program. The workshop also addressed important themes such as assessment, supporting student writing in mathematics, facilitating small group work, and leading a class discussion. The district discovered that this professional development provided teachers a general understanding of the *MathScape* philosophy as well as a specific introduction to materials they would use at the start of the school year.

One teacher, who had assumed her teaching role mid-year, shared her experience of teaching a unit without having had this kind of professional development. She described the experience as daunting and frustrating. She was unsure about pacing. Not having had the chance to work through the activities in the unit with a colleague, it was difficult for her to predict where students might have "a-ha" moments. She had trouble anticipating how things might go, where potential areas of difficulty might lie, and how students might answer questions.

After completing the remainder of the school year, this teacher attended the 3-day summer workshop and noticed a big difference in her experiences in the classroom the following year. She stressed how critical this kind of in-depth professional development was to her success. The opportunity to have dialogue with others, experience the mathematics as a learner, and anticipate how things might work in the classroom made her much more comfortable in taking risks and reaping the rewards of teaching with *MathScape.* A hands-on workshop provides a boost of self-confidence as well as increased content knowledge. It is absolutely essential. Equally important is the continued support and professional development offered by teacher leaders, colleagues, and the curriculum coordinator during the school year.

Mentoring

In addition to workshops, new teachers in the district received the support of a mentor teacher. The new teacher met with the mentor once per week until the teacher was comfortable with requisite planning and instructional issues. Meetings occurred less frequently after this comfort level was reached. These mentor/novice sessions served as "tutorials" for the units that were coming up. The meetings were specific to the new teachers' instructional situations and particular populations of students. These mentors served as local grade-level facilitators who would meet with a grade level group of new teachers four or five times a year. The facilitators focused on different topics related to teaching mathematics. These topics included examination of mathematics content in particular units, strategies for teaching to a range of students in the classroom, integrating technology, and formal and informal assessment.

The district also held monthly department meetings. These provided an opportunity for ongoing teacher support and growth for all teachers, new and veteran alike. One teacher reflected upon how to maximize the effect of these regular but limited opportunities. From her perspective, the "opportunity to share" needs to remain a central part of the agenda to encourage broad participation and make the outcomes of the meetings relevant for teachers. She recommended keeping meetings specific to *MathScape* and topics inspired by the use of *MathScape.* Discussing what activities and tasks worked and what did not enabled teachers to examine their practice. In addition, looking at student work with a structured protocol was considered a valuable exercise that supported the implementation.

CONCLUSION

The story of curricular implementation is a story about change. Change is almost always accompanied by challenges and complications. Change is also a necessary part of any growth process, including the learning of both teachers and students. Through open dialogue and representation of *all* who hold a vested interest in the outcome, change can thrive. An in-depth discussion about the need for change and a process to support change can ensure that no party is marginalized in the process. Approaching change with these ideas in mind may motivate facilitators of change to include many voices from the outset and to check in routinely with the stakeholders in order to evaluate progress along the way. This is the first big step toward success in making and sustaining curricular change.

The story of this particular district is one of a very committed group of teacher leaders. They desired change and educated themselves on how to realize it. They also earned the respect of the community of teachers, administrators, and families. The presence of such people directed by a dedicated curriculum coordinator was an essential ingredient in the initial and ongoing success of their implementation. The lessons they learned and the resulting recommendations outlined below may be useful to other districts embarking on curricular change.

SUMMARY OF LESSONS LEARNED AND RECOMMENDATIONS

1. Work to build enthusiasm around the change while affirming that change presents challenges and takes time. Define manageable and measurable goals for various stages of the implementation and monitor these goals by soliciting input from those involved in effecting the change. Implementing three units each year to phase-in the program was manageable, and acknowledged to teachers and students the careful work required in changing a mathematics program. Doing more than this may overwhelm teachers, parents, and students.
2. Make sure to address the needs of *all* stakeholders in the process. In particular, be sure to include special educators in professional development opportunities. In retrospect, some teachers in the district profiled here felt that there was not enough special education involvement. When this oversight was identified, the district implemented "tutorials" in *MathScape* to provide hands-on experience for special educators. The district also invited special educators to attend the bimonthly workshops with grade-level facilitators.
3. Provide ample and ongoing professional development opportunities that address general aspects of mathematics instruction including diversity of student needs, assessment, writing in mathematics, as well as specific issues unique to the selected instructional materials. Consider ways to tap into the expertise available within your own

district and offer leadership and mentor opportunities to teachers. Make sure that teachers are not isolated in the implementation process. Consider ways to connect teachers with one another to support ongoing reflection, sharing, and problem solving.

4. Take steps to insulate teachers from potential criticism about the curricular change. Some suggestions for doing this include:
    - Encourage teachers to refer questions/concerns to administrators or other designated "specialists" familiar with the curriculum materials and knowledgeable of the adoption process. A teacher's responsibility is to teach students, not to defend to parents the selection of a program. If teachers are put in the awkward position of always defending a district decision, it may negatively influence their attitude toward the change process. That is not to say, of course, that teachers should not have the freedom to discuss these issues with parents. However, teachers should also have the option to say, "Some of your questions might be better addressed by the curriculum coordinator. She could speak to them more fully as she was closely involved in the selection process and can offer the district-wide perspective."
    - Support teachers in communicating with parents. Consider having the curriculum coordinator or teacher leaders facilitate the initial informational meetings for parents. Teachers have enough work in preparing to use the new materials. In shifting the parent communication responsibility to a "central" group, this allows teachers to maintain their classroom focus. It will also result in a consistent message across schools.
    - Dedicate meeting time to create and discuss outlines and handouts for Back to School Night, Open House, or other parent meetings.

AUTHOR NOTE

With special thanks for their contributions to Jullie Rains Snyder, Joan Miceli, and Barbara Scotto.

CHAPTER 11

SUPPORTING CURRICULUM IMPLEMENTATION THROUGH ITS UPS AND DOWNS

Linda A. Tetley
Jefferson City Public Schools (retired), Jefferson City, Missouri

The story of curriculum implementation likely involves as many different themes as there are types of teachers. As teachers, we live in that real world of students and classrooms, not the world of theory. From my own personal experience, the process of implementing new instructional materials was all of these things: exciting, challenging, frustrating, rewarding, worrisome, scary, and often downright thrilling! In this chapter, I relate the implementation of MATH*Thematics* (Billstein & Williamson, 1999a) in a school district I identify as Lakewood. In describing the process, I outline my role as a district consultant and share the challenges and issues raised by teachers I worked with, as they lived their own "stories."

PERSONAL BACKGROUND

Before sharing the implementation of MATH*Thematics* in Lakewood School District, it is worth briefly recounting several key ideas, drawing from my

A Decade of Middle-School Mathematics Curriculum Implementation, pages 107–114

personal experiences in implementing new instructional materials. In my own school district of Jefferson City, Missouri, we received extensive professional development at a summer institute prior to implementing MATH-*Thematics* in the fall. The focus of this work was to become familiar with the underlying concepts of standards-based materials, as well as the specific content of the materials from which we would teach. During our professional development, we learned by working through the materials as if we were students. That is, lessons were introduced to us as a "class." We then were assigned to groups of four to work through the explorations as directed in the textbook, utilizing the suggested questions for discussion and further thought. For example, we might choose a recorder to make our list of outcomes after sorting cards, then discuss patterns we observed before coming to group consensus of a generalization. In another activity we might play a game, then continue to pull out the mathematics embedded in the game by following the suggested sequence in the textbook. We then discussed what our role should be, as teachers, based on what we had encountered in the lesson as students.

This experience of working through the materials as students helped us practice new methods of teaching and learning, as well as make sense of the mathematics content of the lessons. Through this process we practiced our new role as co-investigator and facilitator. We also saw how lessons often take a different turn than planned when an inquiry approach is used. Another important aspect of these professional development sessions was that we had time to study the scope and sequence of the entire middle school mathematics program. We also saw firsthand the integration of content, as well as the spiraling of the mathematical strands, throughout the materials. This would later become important as we thought about where our own students' learning fit into the bigger picture of middle grade mathematics.

The process of implementation was made smoother by this model of professional development. Definitely, the early intense training we received had an impact on our future success. I decided to apply these same methods myself in providing professional development for new users of MATH*Thematics.*

LAKEWOOD SCHOOL DISTRICT

Following my retirement from the classroom and through my subsequent connection with the Show-Me Project, I had the opportunity to work closely with Lakewood School District during their first year of implementing MATH*Thematics.* In my experience, this school district's story is somewhat typical of many others in that they received some initial professional development provided by the publisher, but for various reasons had little follow-up to address problems that developed during that first semester of teaching the new curriculum. I would like to focus on the initial challenges teachers encountered as well as the ways these challenges were addressed and overcome.

Lakewood is a suburban school district with three middle schools—a very typical, moderately sized school district that might be found in many places across the United States. Relative to similar districts, they are a high-achieving district with test scores already well above the state and national averages. Teachers are well educated with over 70% holding at least a Masters degree. Among the district's strongest points are the energy and dedication of its teachers. Another point in the district's favor is strong administrator support for the teachers and for the selected mathematics curriculum. With these conditions, the implementation of MATH*Thematics* seemed poised for success, yet the first year got off to a rocky start.

In the summer prior to beginning the new curriculum, the publisher had provided one full day of in-service. This professional development day involved an overview, a beginning look at grade-level materials along with some management techniques and lesson planning. Teachers had an opportunity to work through several activities. The professional development was as thorough as it could be for a one-day overview. The experience could have been improved significantly with a commitment to several additional days, which would have offered opportunity for a more thorough discussion of the scope and sequence of the materials and more of a focus given to the overall philosophy and pedagogical approach. As is often the case, however, no more time was available prior to beginning of the school year.

Another problem with the initial professional development was, due to time constraints often present in the summer, not all of the teachers were able to attend. No decision was made about how to make up for this lack of preparation, but several teachers volunteered to bring the others along as the year progressed. This support for colleagues was often evident with the teachers in this district, and was an important factor in their success.

EXTRA SUPPORT IN YEAR 1

In October of that first year, the Show-Me Center was able to offer a partnership with this school district to provide additional professional development to the mathematics teachers. The project would support the district, with the frequency and type of support left up to the district's discretion. Later that fall I met with Lakewood's Director of Curriculum and Professional Development to discuss the additional support and services provided by this partnership. This partnership was a commitment that would continue the rest of that school year and all of the next.

My first meeting was not with teachers, but with administrators, and occurred in January during their first year of implementation. The three school principals as well as the Curriculum Director brought a list of concerns for me to address, based on what they were hearing from their teachers and parents. The following is a list of concerns they compiled:

- Parents have questions and concerns
- How are skills, such as operations with fractions, addressed and practiced?
- What about increased reading demands on struggling readers?
- How can we move more quickly through the lessons, yet still address the concepts and skills in a thorough manner?
- How do teachers find the time to plan for this type of lesson and then prepare the needed materials?
- When can teachers include their own special interests and projects?

District administrators pledged to continue to support the teachers in all ways possible. They also agreed to give release time during the school day for in-service and to provide substitute teachers as needed.

A few weeks later, in February, I met for the first time with teachers, holding grade-level meetings, which were attended by all the mathematics teachers. Most of this time was devoted to addressing the immediate concerns held by the teachers and answering their questions based on my own experience as a teacher of these materials. For example, we completed a sample lesson to discuss specific techniques to teach reading of mathematical vocabulary. In this way, we practiced the "teacher learning as a student" model.

We had time for one grade-level activity from each textbook, and I led a discussion of "Best Practices." This session specifically addressed teacher concerns about pacing and skills versus problem-solving issues. The session ended with the teachers completing an evaluation and guidance sheet for the next time we would meet. These proved most revealing! I learned that several major misconceptions existed about how to best use the materials.

Most of the concerns mentioned had arisen due to lack of sufficient knowledge of how the lessons were to be presented and carried out. The misconceptions had simply never been addressed in the brief professional development the previous summer. Since months had now gone by, some teachers had become distressed or at least frustrated by slow progress. Many of these difficulties could have been prevented with earlier intervention. However, in the real world of schools, this intervention simply does not always occur in a timely manner.

Here is the list of concerns generated by the teachers, in order of importance, according to how often the concern was mentioned:

1. Pacing
2. Helping slow readers and slow writers
3. Deciding what material to cover and what may be skipped
4. Keeping students on task and actively involved in their learning
5. Constructing tests and other assessment issues

Teachers also said they could use more information on topics such as sharing ideas and strategies, helping parents think positively about the book, and how the spiraling works in MATH*Thematics.* They universally said they needed more time to prepare for teaching the Explorations, which often involve making game pieces or other hands-on materials, duplicating lab sheets, sorting manipulatives, and assigning homework. They also asked for hints on classroom management and organization, as well as ways to share the work of preparation with other teachers.

After some analysis, I began to realize that a large part of the support these teachers needed was assurance that all of these difficulties were common to the first year of implementation, and that everything on their list could be addressed. They also needed to be reminded that their district was changing the way mathematics was taught and learned, a process that takes time. The MATH*Thematics* curriculum has underlying concepts that are very different from a traditional textbook. Materials are designed to engage students in the processes of problem solving, reasoning, communication, connections, and representation. These essential practices are developed through active investigation and discovery, carefully guided discussions, writing, and teacher questioning. Another difference is that mathematical strands spiral throughout the entire curriculum, rather than being isolated into a chapter or unit. Things would get better the second year and each year thereafter as they applied what they learned about teaching and learning in a new way. These ideas were among the first I reinforced to the teachers.

When we met again in May of Year 1, each of the topics suggested on the teachers' original list of concerns became the area of emphasis for a session. In this way, we all put our heads together to discuss issues and hopefully solve problems. This plan seemed to give the teachers ownership in the whole process. On this particular day, we met as a whole group, rather than by grade level, as the concerns were consistent across the board. Each grade level reported their progress in the book to the others as a way to help in planning for the next year. This concluded the first year of our partnership.

This school district, and the administrators in particular, addressed the changing needs of their teachers by offering support and time to work as colleagues. Early in the year, teachers were unsure what to expect. With experience and time, they came to recognize the questions to ask and were willing to listen to others who had positive experiences to share. A few remained uncomfortable with the new approach to teaching and the changes that were required in their classrooms. Some were not yet on board with the new curriculum and its approach.

However, by the end of the first year, most teachers were talking of the mathematics curriculum in a much more positive way and were looking forward to changes they would make for the second year that would make it even better. The changes in attitude seemed to occur due to giving teachers a sense of direction and giving them an opportunity to collaborate with each other. This commitment on the part of teachers had other benefits. For example, one of the principals pointed out, "Our teachers' focus on professional development and continual improvement transfers to the students' willingness to engage in meaningful and productive classroom work."

YEAR 2: ADJUSTING TO THE CHANGING NEEDS OF TEACHERS

I adjusted my role during Year 2 to meet the changing needs of teachers. Early in the school year, I met with all three grade levels to check in on progress and answer additional questions. I found that each grade level had already met at least once with their colleagues to plan and share ideas. I was pleased to hear the positive stories about how much smoother things were going in this second year. Changes had been made for the better in many areas of instruction and we sat down together by grade level to plan the first semester. We addressed questions about pacing and I scheduled additional visits based on their requests for additional assistance.

For example, one group of teachers scheduled regular meetings with me to work out the pacing of instruction for each upcoming quarter. When we met, we shared success stories, discussed issues that teachers were experiencing, and offered ideas for addressing their concerns. As we planned and worked together, I was able to address the teachers' changing needs in a positive way. The result was that teachers felt good about the instruction they were providing and their students' increased opportunities for learning as they progressed through the curriculum.

Overall, the enthusiasm and expertise of the Lakewood teachers was most impressive! By the end of Year 2, they had taken ownership of the change process and become confident in their use of the MATH*Thematics* materials.

8th-grade teacher: I would say that the professional development has helped me to feel more comfortable and confident, which in turn has built the students' confidence and raised achievement.

8th-grade teacher: Since I know how to teach the curriculum better, the kids seem to be learning and retaining more information.

6th-grade teacher: Meeting with my math cadre has been the most beneficial thing for me since moving to middle school. I love the fact that each teacher feels the same pressure that I do and that Linda is there to soften the blow. We constantly share ideas with each other in the cadre.

Curriculum director for the district: Our teachers are far more vested and supportive of MATH*Thematics* this year. [Linda Tetley] has worked with three buildings to help our teachers "come around" to see the power of this series.

Middle school principal: As I have spoken with math teachers, I have noticed an increased confidence in their instructional toolbox and a decrease in the frustration with the modules taught in Year One. Year Two has been a much better experience for them as they have identified a more suitable pace and have progressed nicely through the modules.

CONCLUSION

There appear to be at least three key factors in the successful implementation of new instructional materials that require a change in pedagogy. Even if the process falters initially, as this example demonstrates, steps may be taken to support teacher change and increase the chance for a successful implementation.

First, the effect of on-going professional development cannot be overstated. This should be intensive both before the implementation and later in follow-up sessions as teachers begin to encounter difficulties. Teachers are empowered by collaborating with each other. They need to know that "we are all in this together" and that everyone will benefit from sharing their knowledge and experiences.

Second, opportunities are needed for teachers to learn about the "big picture" of the program as well as specific content for a particular grade level. The unifying concepts, philosophy, and instructional approach of the curriculum must be understood if teachers are to use the program as the authors intended. I found that possible stumbling blocks to the curriculum may be avoided if sufficient attention is paid to these design principles early on during professional development.

Finally, teachers benefit if they know they have the support of their administrators. In the Lakewood School District this support was obvious. It is too difficult to embrace change if teachers feel they are all alone in their efforts or that they do not have the backing of their administration. When teachers feel secure that administrators support the adoption of new instructional materials, they are more apt to persevere when encountering the challenges inherent in implementing something new. Yet, all involved would probably agree that challenges still remain. Improvement is an ongoing process and this district continues to work to meet their goals.

To conclude, I quote the Director of Curriculum, who showed support to the entire mathematics department time and time again: "This has not been the easiest shift, but easy does not always mean it's in the best interest of our students. I'm confident with where we're headed." I can definitely say that for me, and the teachers with whom I worked, this process of change was worthwhile. Through professional development and effort, colleagues can overcome obstacles so that the story of their school and their students will be one of success.

CHAPTER 12

WHEN THINGS GO WRONG IN CURRICULUM IMPLEMENTATION

Jane O. Swafford
Illinois State University (retired), Normal, Illinois

Cynthia W. Langrall
Illinois State University, Normal, Illinois

The stories presented in the preceding chapters have aptly portrayed the ups and downs of curriculum implementation. In the end, they are all stories of success. But in reality, the implementation of NSF-funded curriculum materials is not always successful, and there is much to be learned from those experiences.

In this chapter, we present the stories of four districts, each representing a distinct image of the implementation process as we have seen it play out in a variety of settings. Although readers might believe they "recognize" one of the districts we portray, we must emphasize that the four narratives are hypothetical cases. They are composites drawn from the experiences of the curriculum developers and patterned after data collected from a series of case studies we conducted over a four-year period to learn more about the implementation process. Our research and experience indicates that there are identifiable factors that can support curriculum implementation, or in

A Decade of Middle-School Mathematics Curriculum Implementation, pages 115–123

their absence, can impede the implementation process. Four broad factors we have identified are: (a) a plan for implementation of the materials; (b) ongoing professional development aimed at the changing needs of teachers; (c) district- and building-level leadership and support; and (d) curriculum alignment with state policies and tests. We present the following stories as examples of what can go wrong during curriculum implementation and to illustrate how the implementation process can be derailed through no fault of the curriculum itself.

STORY 1: THE IMPORTANCE OF AN IMPLEMENTATION PLAN

Our first story is about Lincoln County Public Schools. This district surrounds a large Midwestern city and consists of affluent suburbs, some rural areas, and also urban neighborhoods closer to the city limits in the center of the county. The impetus for using NSF-funded materials in Lincoln County came from the mathematics supervisor, Stephanie Layton. She was new to the district and wanted to see changes in mathematics instruction in Lincoln. She was introduced to the NSF-funded curricula through presentations at the National Council of Supervisors of Mathematics conference. These materials embodied the curriculum changes she envisioned for her new district. When the district began pushing for a gifted mathematics program, Stephanie shared some of the NSF-funded materials with the new gifted coordinator who was looking for more interesting and challenging materials for these students. He liked what he saw and recommended that the NSF-funded materials be used with the gifted classes.

Stephanie saw this initial adoption with the gifted classes as a way to eventually spread the curriculum changes they embodied to all of the mathematics classes in the district's nine middle schools. Her plan was to implement by grade level, using the NSF-funded materials with the grade 6 gifted students in Year 1. However, the assistant superintendent decided at the last minute to push the implementation into the gifted classes at all grades. But no plan was in place to accomplish this.

Things went wrong. First, the number of teachers who needed professional development had tripled. Stephanie scrambled to offer introductory sessions for everyone, but could not address the needs and concerns of teachers at different grade levels. Further, because the materials had been decreed from the assistant superintendent, there was little buy-in from the teachers. Many were totally unfamiliar with the materials and their problem-solving approach. The demands of the materials in terms of preparation time and mathematical knowledge did little to engender the teachers' support and a backlash ensued.

Second, because of the last minute decision to extend the implementation to gifted classes at all grade levels, materials that were ordered for the seventh- and eighth-grade classes did not arrive until half way through the school year. Since the district had a traditional mathematics textbook adopted for the regular classes, it was used initially in these gifted classes, and in some of cases, was never fully abandoned after the arrival of the NSF-funded materials.

Third, there was no effort to develop a customized sequencing and pacing plan for the year. Hence, many teachers began with unit one and found they could only cover about half of the units that first year. Others, realizing that they were not going to be able to cover all of units, started picking and choosing among the units and lessons within units. With no guidance, the result was a curriculum that lacked cohesiveness within and across grade levels and one that poorly matched the state standards and tests. Without any plan to phase in the materials or to align them with the state standards and tests, the implementation in Lincoln County ran into trouble.

In Year 2, only some of the teachers of gifted classes were using the NSF-funded materials as their primary text. All teachers supplemented from the traditional textbook, mostly for extra practice but also for problem-solving and application activities. In some of the middle schools, the curriculum materials were no longer being used with any of the gifted classes. Stephanie lost hope that the materials would be adopted into regular classes once parents and teachers saw that "different" materials were being used with the gifted students. At the same time, the administration began to put increased pressure on schools to maintain or exceed the high ranking the district had received on the state testing program. It was difficult for teachers to see how the NSF-funded materials were addressing all of the state standards. Frustration reigned. By Year 3, the NSF-funded materials had all but been abandoned and the district had adopted a new traditional middle school textbook series that was used in all classes.

STORY 2: THE IMPORTANCE OF PROFESSIONAL DEVELOPMENT

Ciudad Grande Public Schools, a large urban school district in the Southwest, did not make the mistake Lincoln County did. The curriculum supervisor, Lynda Martinez, knew from experience that she would need a plan for phasing in the new NSF-funded materials. For the first year of implementation, she formed a committee of middle school mathematics teachers who identified specific units and lessons at each grade level that closely matched the district's mathematics framework. Realistically, Lynda did not expect teachers to be able to complete all the units in their first year using the new materials. So a plan was developed to phase in the materials using core units in Year 1. In subsequent years, she expected teachers to be able to cover the entire curriculum.

In addition to developing a plan to phase in the materials, Lynda also mounted a large professional development program for all teachers in Year 1. In cooperation with the local university, the district offered an intensive one-week summer course for anyone who was interested. The district also provided three days of professional development by grade level at the beginning of the year. Additionally, there were four professional development days built into the calendar for quarterly grade-level workshops. During these sessions teachers worked through upcoming units to organize and plan for instruction. The district also sponsored four optional Saturday workshops, a workshop for lead teachers, and a half-day orientation to the materials for building principals.

Although there were multiple and varied professional development opportunities available, most of them were voluntary. Only the three-day grade-level inservice at the beginning of the year was required. In this large district, the building principals had a lot of autonomy. Some principals sent all of their mathematics teachers to the quarterly grade-level workshops, some principals sent one teacher for each grade level, and some principals used the four professional development days to address other priorities. However, those teachers who participated in the full range of professional development activities found them very beneficial in implementing the new materials.

Lynda had been able to offer a robust professional development program in Year 1. However, the district lent her less support in subsequent years. In Year 2, although the summer course and quarterly workshops were available, the three days of professional development at the beginning of the year addressed literacy, not mathematics. As do many urban school districts, Ciudad Grande experiences frequent teacher turnover. New teachers, most of whom were hired just before the start of the school year, were expected to buddy up with a teacher who had used the materials in Year 1. The quarterly after-school grade-level work sessions continued to be opportunities for teachers to work through upcoming units to organize and plan for instruction. However, since many of the teachers had already taught with the materials, participation in these voluntary sessions was not strong. At the same time, some of the teachers–specifically those who were either elementary certified, teaching out of field, or participating in an alternative certification program while teaching–were experiencing difficulty with some of the mathematical content in the materials. Their need to "brush up" on their mathematics knowledge went unaddressed.

Although the NSF-funded curriculum was launched with strong support and enthusiasm, Ciudad Grande failed to provide ongoing professional development aimed at the changing needs of their teachers. Using the mind set of a traditional textbook adoption, they believed adequate inservice in Year 1 was all that was needed. But without ongoing and appropriate professional development, many teachers simply gave up on the curriculum. Over time the number of untrained teachers exceeded the number of teachers who had participated in professional development. By Year 4, only one or two teachers per building were still using the NSF-funded materials. The others had reverted to their old traditional textbook, which they were more comfortable with, and were looking forward to the next adoption cycle.

STORY 3: THE ROLE OF LEADERSHIP

The story of Milton Unified Schools illustrates the importance of strong leadership at both the district and building levels. Milton is a small town in the northeast, close to a major metropolitan area. Historically, the community's economy had been dominated by manufacturing and residents had been accustomed to receiving premium pay for unskilled or semi-skilled work. Although valued, education had not always received the highest priority because it was not essential to earning a comfortable living. However, job requirements were changing in Milton and employment for unskilled labor was rapidly diminishing. Thus the community was realizing the need to move from an industrial economy to an information economy, and accordingly, to revitalize and improve Milton's educational system.

Mathematics education was a content area targeted for improvement and members of the district curriculum committee decided that the middle school was the place to start. The NSF-funded mathematics curriculum program that was adopted had received strong support from two teachers with close connections to the curriculum developers. These teachers, Bart Sanders and Donna Aspen, had been involved in piloting the materials during the development of the curriculum and had provided extensive feedback to the curriculum developers and the publishers. Needless to say, Bart and Donna championed the adoption and were able to provide assistance to the teachers in their respective schools during the implementation process. They also helped the district's professional development coordinator, Marv Henderson, and the curriculum coordinator, Clare Holmes, develop and implement an extensive professional development program for all middle school teachers during the first year of implementation. In subsequent years, the number of professional development days for curriculum implementation gradually diminished and by Year 4, only new teachers received professional development for using the materials.

Throughout the first three years of the implementation process, teachers perceived Marv and Clare as strong supporters of the curriculum. Teachers appreciated their efforts to photocopy materials, purchase supplies, communicate with parents, introduce principals to best pedagogical practices in mathematics, and align district assessments with the curriculum materials. As one teacher commented, "If there is anything we need, Marv will make sure we get it!" The leadership provided by Marv and Clare was a vital component of the implementation of the NSF-funded curriculum in Milton.

The implementation process was not without its problems, however. Throughout the process, the grade 8 teachers were reticent and not willing to implement the curriculum fully with their students. Some of these teachers made disparaging remarks about the curriculum and tried to dissuade grade 7 teachers from using the materials. For the most part, however, teachers in grades 6 and 7 stayed the course and realized the benefits of the new program, particularly in terms of students' problem solving and critical thinking. Their resolve was bolstered by the support of a few strong teachers, like Bart and Donna, perceived as leaders in the district. As a result of the unrest initiated by the grade 8 teachers, the principal at one school set clear expectations that all teachers use the NSF-funded curriculum materials as the basis of instruction, and in another school the role of the lead teacher was expressly focused on supporting the implementation process.

Even with these efforts, obstacles were able to take root and grow mainly due to changes in the leadership in the Milton system. By Year 3, professional development sessions for mathematics were waning as the district had shifted its focus to a new literacy initiative. Marv Henderson had accepted a job outside of the district and his position as coordinator of professional development was eliminated. Donna Apsen, a respected teacher leader and strong supporter of the curriculum materials, left the middle school to teach in the elementary grades. By the end of Year 4, the principals at two of the middle schools had retired, lead teacher positions had been abolished, and Clare Holmes, the curriculum coordinator, had accepted a position in a neighboring district. The new curriculum coordinator was not familiar with the NSF-funded curriculum materials and showed no interest in continuing to support the implementation. There was no one in the district willing to speak up about why the materials had been adopted or to promote the success that had been realized by teachers and students using these materials. Under the direction of the new curriculum coordinator, a new traditional textbook was subsequently adopted.

STORY 4: THE INFLUENCE OF STATE POLICIES

Even with strong leadership, a solid implementation plan, and adequate professional development, things can still go wrong. This is illustrated by the Cedar Creek School District implementation story. Cedar Creek is a bedroom community for a large city in the southern part of the country. It serves a diverse population of students from urban poor to suburban middle class. In this district, the middle school mathematics curriculum had been quite traditional, but the high schools had become more progressive and were implementing a combination of innovative and traditional materials. The high school mathematics teachers campaigned for a more innovative curriculum at the middle school level to better prepare students for the transition to grade 9. To this end, the middle school curriculum committee agreed on one of the NSF-funded programs included on the state textbook adoption list.

Rita Ivins, the district coordinator of professional development, initiated a well-designed scheme for implementing the new curriculum materials. She developed an implementation plan that matched the topics and pacing of the materials to the state standards. With the assistance of mathematics educators from a local university and the curriculum developers, she planned and conducted a professional development program that had both initial training for all middle school mathematics teachers as well as ongoing support for the first four years of the implementation. Additionally, she cultivated a cadre of teacher leaders to provide building level support for teachers. With Rita's leadership at the district level, strong support from the superintendent and the middle school principals, and the onsite assistance of the lead teachers in each school, the implementation was poised for success.

By the end of Year 1, teacher buy-in across the grades was generally high. Most teachers reported using the curriculum materials as the basis for instruction although all teachers supplemented the curriculum materials with drill and practice activities. By the end of Year 3, the overall achievement test scores for middle school students were on the rise. All in all, the implementation was going well.

In Year 4, significant changes at the state level had dramatic effects on the implementation. First, the state mathematics standards were revised placing greater emphasis on procedural competence, increasing the number of outcomes at each grade level, and shifting some topics to earlier grades. The state assessment was also revised to reflect the new state standards. As a result, the sequencing of topics in the curriculum materials was no longer a good match with the state standards or assessment. However, an even greater barrier to continued implementation was that the NSF-funded curriculum program was removed from the state list of approved textbooks. Only textbooks meeting the new state standards were approved for adoption. Further, districts not using one of the approved textbooks in the upcoming school year would be penalized by a reduced allocation per student. Hence, Cedar Creek was forced to adopt another middle school textbook, one traditional in its pedagogical approach.

Although the implementation of the NSF-funded curriculum officially ended in Year 4, Rita Ivins remained optimistic that mathematics reform, however, would not die. The new textbook was admittedly procedural in focus, with each lesson presenting a rule, examples, and exercises with a few word problems at the end of the exercises. But from Rita's perspective, this was not a disaster; instruction was still be the responsibility of the teacher. It would be up to the teacher to develop students' conceptual understanding through hands-on activities, problem solving, and investigations. The NSF-funded curriculum materials remained in the classrooms with the hope that they would be used to supplement the new textbooks. Rita believed that the implementation experience for Cedar Creek teachers and students had been a valuable one. Her hope was that the spirit of the NSF-funded curriculum materials, and even some of the activities, would live on in many of Cedar Creek's middle school classrooms.

CONCLUSION

Successful implementation is a highly complex process. It takes dedicated and skilled leadership to plan and execute a successful implementation. It also requires adequate resources, both human and financial. There must be talented people to conduct professional development and the budgetary commitment to support them. And there must be strong teacher buy-in to achieve substantial change in the way mathematics is taught and learned in our culture. The chapters by Sharp, Den Otter, Fagan and Weinstein, and Tetley (this volume) illustrate the complexities of the implementation process and highlight the rewards of a successful implementation.

In contrast, the stories we have presented illuminate circumstances that can derail curriculum implementation. Although they may seem fatalistic, we do not mean to imply that an implementation that starts on the wrong track necessarily ends in a train wreck. Linda Tetley's work in the Lakewood School District (Chapter 11) shows how a problematic implementation can be rescued through continued support and targeted professional development. Rather, we intend these cases to be provocative in posing issues that many districts are likely to face when implementing NSF-funded curriculum materials. We want the stories presented in this chapter to raise questions in the mind of the reader about what went wrong in these districts and how some of the problems might have been prevented or handled differently. We hope the cases of these four districts serve as catalysts for discussion; and toward that goal, we offer the following commentary.

In Story 1, the NSF-funded materials never had a chance. In effect, the lack of a systematic plan for implementation kept the envisioned changes in the curriculum from ever being achieved. But would a multiyear phase-in of the curriculum materials with gifted students have been enough? How could the district have expanded the implementation from gifted classrooms to regular classrooms?

In Story 2, the lack of ongoing and appropriate professional development undermined the implementation. Would it have been enough for the district to continue to dedicate the beginning of the year and quarterly professional development days to mathematics? Or would frequent teacher turnover and the use of non-certified teachers challenge even the most robust mathematics professional development program?

In Story 3, the loss of leadership and advocacy for the NSF-funded curriculum materials led to the demise of the implementation. Strong, sustained leadership at the district and building levels is an important component for the successful implementation of any curriculum program, but it is critical for the NSF-funded curriculum materials. However, one wonders whether things would have turned out differently if leadership in the district had remained stable. How could the district have addressed the lack of commitment to the implementation on the part of Grade 8 teachers?

Finally, in Story 4, the implementation was curtailed because of intervening state policies. But after three years of solid use, the problem solving, hands-on approach of the NSF-funded curriculum had become standard practice for many of the teachers in the district. Was this change in pedagogy enough? Is it reasonable to expect teachers to maintain an inquiry approach to instruction while using a traditional textbook? What difficulties might they face in using the NSF-funded materials as a supplement?

We take the stance that even in districts where "things went wrong," the attempt to implement NSF-funded curriculum materials cannot be considered a failure. Many teachers were exposed to a new way of thinking about the teaching and learning of mathematics. And although they may not have embraced this new approach, their routine of "demonstrate, then drill" was challenged. At the very least, these implementation efforts presented teachers an invitation—an invitation to doubt. It may not be enough, but it is a start.

CHAPTER 13

ISSUES OF IMPLEMENTATION

Making Mathematics Accessible to All Learners

Jennifer M. Bay-Williams
University of Louisville, Louisville, Kentucky

Rebecca McGraw
University of Arizona, Tucson, Arizona

Because the curriculum programs funded by the National Science Foundation (NSF) were developed to reflect the vision of the National Council of Teachers of Mathematics (NCTM) *Curriculum and Evaluation Standards for School Mathematics* (NCTM, 1989), lessons in the curricula are characterized by an increased use of problem solving, reasoning, and communication. There is an expectation that students will be able to do such processes as analyzing the problem, selecting solution strategies, making generalizations, and explaining (in writing and orally) how and why they solved a problem in a particular manner. Collectively these expectations mean more higher-level thinking and more reading, writing, speaking, and listening on the part of students. It is not surprising that questions have been raised about how to effectively teach such a lesson with the range of learners in any given classroom. For example, how can English language learners be appropri-

A Decade of Middle-School Mathematics Curriculum Implementation, pages 125–137

ately engaged without becoming frustrated? How do curricula meet the needs of struggling learners and also challenge the gifted students?

The NSF-funded curricula are designed to meet the needs of all learners in the classroom, from struggling learner to gifted learner. The use of many contexts, encouraging different approaches to solving a given problem, and small-group learning opportunities are a few of the ways that provide access to challenging mathematics for a range of learners. Still, implementing a curriculum that is demanding in terms of language and mathematics has shifted the type of support that students need to be successful. This chapter characterizes two student-learning issues that raise concerns during implementation: language and literacy, and meeting the needs of diverse learners. We present successful strategies, gathered from research and teacher experience, for both of these issues.

LANGUAGE AND LITERACY

Traditionally, middle school mathematics texts have offered little opportunity for students to comprehend or use language. The story problems that would follow a set of exercises were solved in the same manner as the exercises, so students could often successfully solve these tasks by extracting the numbers without reading the text. In contrast, the NSF-funded curriculum projects incorporate the four areas of language development: reading, writing, listening, and speaking. Teachers implementing an NSF-sponsored curriculum have questioned both the appropriateness of having a language-rich mathematics program and the strategies to support struggling readers and writers. The use of language in the learning of mathematics is grounded in a solid research base. Communication and representation, two of the NCTM process standards (NCTM, 1989, 2000) underscore the need for students to communicate mathematically. Knowing to incorporate language and knowing *how to* incorporate language are two different things. We briefly summarize the typical challenges that teachers encounter related to language use, and then offer strategies for supporting the use of language in mathematics instruction.

Reading Comprehension

One major hurdle for successful implementation of a language-rich program is related to reading comprehension. Typical to the curricula are lengthy, contextual problems for students to explore. Teachers and parents may worry that the reading will impede a student's opportunity to do mathematics. Students sometimes balk at the quantity of the text to be read and have trouble pulling the mathematical task from the paragraph. Students state that they do not know what the question is asking or that they do not know what to do. This is a particular concern when the task is one of the homework problems and students do not have access to other students or the teacher and do not know how to proceed. But comprehension in reading mathematics requires awareness of how mathematics text varies from text in other domains. The text in mathematics is very dense, containing more ideas per sentence than any other type of text (Kenney, 2005). In addition, text can contain words as well as numeric and non-numeric symbols. Graphs, charts, and illustrations, that are intended to support the text, must be interpreted by the reader.

The first step in Polya's (1973) four-step problem-solving heuristic, understand the problem, pertains to reading comprehension. Several approaches can be used to support students' reading comprehension. For reading comprehension, teachers have often used the following:

1. Through informal questioning, determine if the context is familiar to students. If not, provide background on the context.
2. Be sure students understand the broader goal of the lesson. This is not the same as telling students what they are going to discover or how to do the mathematics they will encounter, but having some direction is important in helping students with reading comprehension.
3. Model the process of interpretation through "think aloud."
4. Ask students what difficulties they are having in understanding the problem.
5. Restate the problem in an interactive manner. For example, one partner can state what the problem is about and the other partner can rephrase the question. Another possibility is to rewrite the problem in every-day wording using short sentences.
6. Provide visual supports to represent the problem context, concept, or vocabulary.

Because the syntax of mathematics story problems is different than everyday language, it is important for students to have opportunities to make sense of how to read mathematics.

Writing

Teachers struggle to get students to explain how they solved a problem or explain a mathematical concept, especially in writing. Each of the curriculum projects includes opportunities within a lesson and at points through the unit for students to reflect on their learning in writing. The precision required in communicating mathematically increases the difficulty of this task. Those learners with learning disabilities in the area of language, in particular, are challenged in explaining their thought processes in writing. Pugalee (2005) argued that writing is important in supporting students' metacognitive processing and improves student understanding:

> As students write, they manipulate, integrate, and restructure knowledge through using and reflecting on what they know and believe. This process facilitates the development of meaningful understanding: the more a student works with ideas and concepts, the more likely those ideas and concepts will be understood and remembered. (p. xii)

Writing in mathematics is something that must be taught to students through guidance and feedback. A number of strategies have been used by teachers to support writing in mathematics. Some teachers have used the Polya four-step, problem-solving heuristic as a guide, asking students to first state the problem, tell what approach they used, what solution they got, and connecting that solution back to the original problem. Written work often follows the classroom discussion. Having students summarize the important concepts verbally at the conclusion of the discussion can support students' individual written summaries. If the writing is getting tedious for students, there are many graphic organizers available commercially (McIntosh, 1997). The written summaries of lessons or of problems, if kept in a math journal, not only serve to help a student synthesize and solidify what they have learned, but become a resource for students when doing homework and in future investigations.

What is essential in supporting writing is to provide feedback regarding the expectations for the written product. Teachers have used peer review of writing as a vehicle for getting more and better writing from students while reducing the amount of papers to review. Another strategy is to have the entire class critique a student's response (or a hypothetical response) to engage students in a discussion of what is expected in the written product. Sjoberg, Slavit, and Coon (2004) created a rubric for assessing student writing. They found that using three specific prompts increased students' mathematical communication markedly. Those prompts, in brief, were to (1) explain, (2) connect to other mathematics, and (3) relate to real world.

Van Boening (1999) reflected on how her students' writing experiences in a standards-based curriculum supported both their mathematical understanding and their language development:

> In addition to becoming better problem solvers, my students are becoming better writers. They are required to give reasons for their answers and to explain how they arrive at solutions. Although they do not always enjoy explaining themselves, they are getting better at writing their strategies and explana-

> tions. This practice in writing and verbalization is important in developing students' thinking and communication skills. (p. 32)

Van Boening's reflections on the relationship between language and mathematics learning are well-grounded in constructivist learning theory, which contends that learning is an active process whereby a learner uses prior knowledge and experience to make sense of new information. Vygotsky calls this process *semiotic mediation*, which is use of language, diagrams, mathematical symbols, and body language as tools that enable learning of a new concept to occur (Forman & McPhail, 1993).

Discourse

A third challenge, and perhaps the most significant, is facilitating mathematical discourse around the important concepts of the lesson. The instructional models of the middle school standards-based curricula include an opportunity at the end of an exploration for students to explain how they approached the problem, listen to how others solved the problem, and to explain connections to other mathematics and to the world. This is a challenge for teachers first in allotting the necessary time for students to be able to share and compare strategies and solutions. In addition, facilitating such a discussion requires constraint and skill on the part of the teacher: deciding when to guide students and when to wait, when to ask a student for further elaboration, and when to elicit input from others. Even when teachers are questioning students, they may not be focusing students on the important mathematical concepts. Engaging all students in this discussion is a challenge, as some students are more willing and more able to articulate ideas than others. Finally, establishing an environment where students feel safe to share ideas, whether they turn out to be valid or not, is not easy to do in middle school.

Kenney (2005) characterized three types of discourse. The first is "traditional," in which the teacher and several students talk. Second is a "probing" discussion, where the teacher is still generating the questions, but the questions are more open-ended. Third, is "discourse-rich," where students are working together toward mathematical understanding. This third form of discussion is the goal for students who are learning with the NSF-funded curricula projects. Facilitating a classroom discussion around a mathematical task is a skill that improves over time. First, developing an understanding of what the discourse is to look like is important. Teachers have found it particularly useful to observe teachers who are adept at engaging students in sharing and listening to each other's ideas. Useful methods for bringing out student thinking during discussion include (1) distributing overhead transparencies and pens ahead of time for students to write on–thereby helping students reflect on their thinking and how to best communicate it, (2) collecting students' solutions at the end of the class period, and using the solutions to prepare for discussion of the following day, (3) establishing norms and expectations for listening and speaking during discussions, (4) planning the order of speakers so that mathematics concepts are developed, and (5) using strategies such as "wait time" and "think- pair-share" that give students time to formulate their thoughts.

Instructional Strategies to Support English Language Learners

English language learners (or ELLs), like all learners, need to make sense of the mathematics they are learning. A key characteristic of standards-based curricula is that they develop a deeper understanding of the concepts, bridging concrete experiences with abstract representations. Consider the problem $1\frac{3}{4} \div \frac{1}{2} =$ ___ . Students can solve this problem by applying the "invert and multiply" rule. This requires little use of language, and can be performed without conceptual understanding of division of fractions. A student who understands division of fractions should be able to model it visually, describe a situation that fits it, and explain why an algorithm works. The standards-based curricula are designed to develop this comprehensive understanding of middle school mathematics topics. Developing this meaning is language dependent. When teachers feel that the language (e.g., reading, writing, speaking, and listening) is an impediment to ELLs, they may move too quickly to the symbolic representation before the conceptual understanding is developed, which limits students' understanding of mathematics and, therefore, their access to higher mathematics (Garrison & Mora, 1999). Mathematics is often believed to be a universal language, a language of symbols that can be interpreted across cultures. However, symbols are not universal. For example, in Mexico, textbooks may record the angle symbol ($\angle$) as a carat (^) over the variable. An ELL may not recognize the angle symbol and in addition might confuse it with the "less than" symbol (Perkins & Flores, 2002). Second, symbols do not have meaning in and of themselves. The concept of "less than or equal to" is attached to the symbol $\leq$ and that concept does not exist separate from that language. As students enter into algebra, they need to understand the meaning of many symbols. Finally, symbolic representations are more abstract than other representations, so more concrete representations should precede symbolic representations.

The NSF-funded curricula embrace the NCTM (2000) Process Standards (problem solving, reasoning and proof, communication, connections, and representations). Inquiry, student-student interactions (pairs and small groups), discussions, and alternative assessments are at the heart of these programs. A glance at any one of the student books reveals the increased use of language in learning mathematics. Teachers using these programs have struggled with students' ability to understand the story problems, in particular the vocabulary. In addition, some of the contexts are not familiar to recent immigrants. Another challenge for teachers of ELLs is how to structure cooperative learning so that the English language learner can participate in the mathematics discussion. As teachers try to limit the linguistic load, they are challenged to adapt the investigation of the day in a way that honors the goals of the lesson. Time is also a challenge, as those moving between two languages as they learn often need additional time to read and write.

The NCTM (2000) process standards support the learning of a student who is also learning English or who is not familiar with particular aspects of U.S. culture (Echevarria, Vogt, & Short, 2004). Khisty (1997) argued that creating effective learning for language-minority students involves integrating principles of bilingual education with content instruction. Using a language-rich curriculum, such as the NSF-funded curriculum projects, is one vehicle for supporting students' mathematical development *and* their language development. Consider, for example, a practice common to these curricula of using student-invented strategies to solve problems and sharing those strategies at the conclusion of the lesson. Encouraging students to choose a strategy enables them to make use of strategies from other cultures, and explaining their strategy and listening to others provides opportunities to develop their language skills. Providing opportunities to talk in a non-threatening environment, using contexts so that students can glean meaning of new words, and using visuals are among the strategies supported in the literature about English language learners. A language-rich environment is necessary, but not sufficient, to support ELLs in a mathematics classroom.

Many strategies to support English language learners exist. Here we share the most commonly used strategies that support learning in the NSF-funded mathematics curricula.

Provide opportunities to speak, write, read, and listen. Maximize opportunities for students to do all of these in non-threatening settings. ELLs can participate in discussions where they grapple with important mathematical content as they are learning language (Moschkovich, 1999; Khisty, 1997). These opportunities need to be structured in a way that an English language learner is able to participate. Providing wait time or time to discuss with a partner, for example, will increase participation in whole class discussions. Providing sentence stems or graphic organizers will increase the quality of writing. Students can take turns reading and then paraphrasing with partners.

Explicitly teach vocabulary as it is encountered in a context. If the term is a mathematical one that must be learned permanently, then it is important to have students add the term to a journal, including illustrations and translations.

It might also be displayed on a word wall in the classroom. If the term is only needed for a particular lesson (i.e., related to the context, but not the mathematics), then it can be written where it is easily visible for the lesson. Williams (2001) suggested that teachers draw connections between English and Spanish cognates, use visual imagery, and emphasize written as well as oral skills. Constantly be on the look-out for words or phrases that take on a different meaning in a mathematics class than otherwise—synonyms, homonyms, and terms unique to mathematics. Pointing out differences and similarities in such words can eliminate confusion for English language learners.

Use comprehensible speech. Comprehensible speech means that words are carefully chosen, reducing unnecessary verbiage and replacing unfamiliar words with more familiar words (Herman & Abedi, 2004). This process does not mean to removing the context of a problem. The context provides necessary support for understanding the mathematics. One must determine what language is needed to keep the task intact, and what language is only going to bog down English language learners.

Build students' background knowledge. Rather than simplify a task or remove language, a tendency in accommodating English language learners that does more harm than good, take time to build up to the task at hand. Spending a short time at the start of a lesson to explore some prerequisite skills provides a scaffold into the mathematical goals for the day. Building background includes a focus on the context and the vocabulary needed to learn the mathematics. Visuals, every-day objects, and examples can help build background.

Encourage use of native language during class work and on assessments. It is important for students' confidence and for their capacity to do mathematics that they be allowed to use whatever language best facilitates their thinking about the mathematics. Illustrations are also important to assess what students know. Multiple forms of assessment support students' development of oral and written skills and provide teachers with the means to assess students' progress in language acquisition and mathematical understanding.

Create a cultural/linguistic biography for each student. English language learners come from a range of backgrounds, and may vary greatly in terms of mathematical knowledge and prior experience. Moschkovich (1999) provided a useful image of two very different students:

> [One student} is a recent arrival, has missed three years of school in her home country, and is orally fluent in Spanish but not proficient in writing or reading... [Another student] is a recent arrival, has not missed any school in her home country, and has completed a year of algebra in Spanish. (p. 5)

The needs of each of these students would be different than those of a student who immigrated to the United States several years ago, or a student whose entire school experience had been in English. Using the strategies mentioned here (e.g., use of native language, building background, and comprehensible speech) are ways to distinguish if a student is struggling with the language or with the mathematics and then provide the appropriate support.

ELLs are best supported in environments that are language-rich and provide concrete, visual models for developing concepts. The NSF-curricula meet the needs of such learners better than traditional, skill-oriented curricula. The teacher in such a classroom must review the lesson with an eye on the learner and build in the support that is needed without reducing the mathematical significance of the lesson.

MATHEMATICS FOR ALL

The NSF-funded curricula were designed for all students, including low, middle, and high achievers, exceptional students (e.g., students with special needs, mathematically gifted students), and ELLs (discussed in the previous section). However, a perception persists that the curricula may not be appropriate for all students. Some feel that the curricular focus on mathematical reasoning and connections within and across concepts, as well as the development of procedural fluency, is appropriate for only gifted and/or high-achieving students. Conversely, individuals may incorrectly conclude that exploration and concept development are needed only for students who have not been successful with other curricula. Both perceptions are problematic, as all students deserve the opportunity to develop procedural fluency and conceptual understanding, and the ability to reason and communicate about mathematics. In addition, the NSF-funded curricula are intended to build upon the intellectual resources of all students. Thus learners benefit from heterogeneous classrooms where multiple and varied ways of thinking and communicating can be drawn out and used to support and foster learning.

Specific areas of concern related to NSF-funded curricula and the goal of "mathematics for all" addressed in this section include accommodating students with special needs and providing appropriate instruction for mathematically gifted students. Readers should note that strategies suggested for addressing one concern frequently address other concerns as well. In addition, general strategies for realizing the potential of heterogeneous, diverse student groups are shared in the final section.

Facilitating the Mathematical Development of Students with Special Needs

Students with learning disabilities may also have a disability in the area of language. These students are most successful (or less frustrated) in solving simple computational tasks. One challenge for teachers is to convince these students, their parents, and their special education teachers, that the move to a language-rich program can better meet the needs of the child. Another challenge that students with special needs encounter is that the NSF-funded curricula offer far less skill practice and repetition. The curricula also have projects or bigger problems to solve over time or one task to work on for an entire period. These tasks can seem overwhelming to some learners. Another challenge is the instructional approach, at least initially. Some students with special needs are accustomed to having steps explained to them and lack the confidence to design their own approach to solving a problem. All learners benefit from a conceptual, concrete approach to instruction. The fact that ideas develop over time in these curricula, rather than a new skill being introduced daily, allows for development of meaningful mathematics for all learners. Here we share some strategies for meeting the needs of students with learning disabilities in this environment.

Use limited instructional time wisely. The NSF-funded curricula focus on the development of conceptual understanding of key topics, replace drill and practice with problem solving and analysis, and encourage the development of multiple solution strategies. All students should be provided opportunities to experience this process. For some students with special needs, this may mean working on fewer problems, but working them through completely—including opportunities for reflection by discussing strategies and comparing solutions.

Embed skill development within instruction. Students with learning disabilities may come to middle school without having mastered algorithms for basic arithmetic, such as addition or subtraction (Woodward & Howard, 1994). Teaching basic skills to these students in ways that are disconnected from other topics and conceptual underpinnings does not lead to mastery or retention. One strategy that has been used with English language learners and with students with learning disabilities is to preteach skills that will be used in the lesson. While this seems to put students ahead, this practice detracts from their learning in that it takes time away from their conceptual development, which should precede skill development. Students are better served when remedial skill development is embedded within instruction. As they work through the curriculum, teachers can decide when and to what extent to focus on skill development. Questions such as "Is developing the skill to do this calculation a student learning goal for this activity?" and "If students spend time working on skill development during this activity, what other learning opportunities will they miss?" can help teachers to make such decisions.

Use technology to support conceptual development. Difficulty in mastering particular skills need not cause students to miss opportunities to develop understanding, and reason and communicate. If a teacher determines that the goal for a lesson is not to apply a particular skill, but instead to develop a concept, and the skill is not yet mastered by a student, then calculators should be provided. Calculators and other technologies can make mathematics more accessible—allowing the answer to a calculation to become a building block (rather than a road block) to higher-level thinking. All students deserve opportunities to build conceptual understanding, make connections across mathematical content, and reason and communicate mathematically, regardless of deficiencies in the mastery of basic skills.

Challenging Mathematically Promising Students

Parents of mathematically promising students want to know if the NSF-funded curriculum projects are sufficiently challenging for their child. Part of this issue is due to the fact that it is harder to see at a glance what mathematics is being covered, as it is easy to do in traditional texts by glancing at the textbook's table of contents. In addition, it is easier to look at the skills a student is learning than to look at the depth of understanding a student is going to develop. The issue to communicate to parents with mathematically promising students is that the depth and breadth in these curricula is well beyond what they would receive in a mainstream textbook and that it truly broadens and deepens their child's understanding, rather than just accelerating their learning.

Focus on depth and reasoning. Students with mathematical promise are not well-served by programs that simply accelerate them through the curriculum. Instead, instruction should provide opportunities for exploring the depth and complexity of mathematics. Because the NSF-funded curricula are based in the exploration of non-routine problems, mathematically promising students can experience the richness and connectedness of mathematics. They can begin to develop the "habits of mind" used by mathematicians themselves, including searching for patterns, experimenting, describing, inventing, visualizing, and conjecturing (Cuoco, Goldenberg, & Mark, 1996).

Use differentiated instruction. Suggestions for curriculum modifications for mathematically promising students in heterogeneous classrooms include the use of (1) extensions and applications, (2) open-ended questions, and (3) self-selected problems (Reed, 2004). Extensions, applications, and open-ended questions can be found throughout the NSF-funded curricula. In some cases, students can go directly to the extension and in other cases they can do the extension in addition to the exploration. The fact that the tasks themselves are open-ended lends to differentiating instruction. Talented students are likely to use more sophisticated strategies than their peers. When they share their strategies or observations, others can see other ways to approach the problem, deepening the learning for everyone.

Realizing the Potential of Heterogeneous, Diverse Student Groups

The NSF-funded curriculum projects are designed to meet a range of learners in a heterogeneous setting. In addition to the strategies already offered for ELLs, students with special needs, and talented students, we share strategies that address the range of learners in the classroom.

Build upon what students know and can do. Characteristic of the NSF-funded curricula are investigations that have multiple entry points. In other words, the tasks can be approached in various ways so a person with a different set of prerequisite skills can still find a way to solve the problem. This allows students to build upon what they know and can do. First, they are allowed to build on their knowledge and ways of thinking as they begin to investigate problems. Second, they can engage in the process of "doing mathematics;" that is, making, testing, and proving conjectures. Third, they can communicate their solutions and solution strategies to peers who have also been working on the same mathematics—or at least the same "big ideas." Lastly, during whole class discussions, they benefit from thinking about other students' ways of thinking, which may be very different from their own.

Explore various solution paths and strategies. The NSF-funded curricula are built on the premise that "thinking about multiple solutions can promote deeper thinking about mathematics" (Burrill, 1998). The problems that form the basis of these curricula contain multiple entry points, and allow students to follow paths and develop solutions that make sense to them. The differentiation in instruction needed to work successfully with heterogeneous groups of students is contained within the curricula, and group work provides a basis for explorations of various solution strategies during whole-class discussions.

Appreciate creativity and divergent thinking. The norms and expectations teachers establish in their classrooms can have a tremendous influence on mathematically promising students. When creativity, invention, and alternative ways of thinking are valued, students become more confident in their abilities. Rather than expecting students to complete problems in the same way as one another, teachers should support students in exploring multiple ways of thinking, and help them see connections across methods. When students realize that their ways of thinking can inform the understanding of the class as a whole, they become more willing to pursue ideas and then share them with others. The structure of the NSF-funded curricula supports creative and divergent thinking. For additional information about fostering mathematical talent, readers should see Sheffield (1999).

Make connections across multiple representations. Words, pictures, tables, graphs, and symbols are used to create and convey meaning in mathematics. Students differ with respect to their preference for, and comfort level with, these representations. The NSF-funded curricula are structured so as to create opportunities for students to work with multiple representations. Teachers can help students make connections across representations by asking them to use a second form to express their thinking, by including several representations of an idea during whole-class discussions, and by focusing students' attention on the ways various representations of ideas are connected to one another.

SUMMARY

The NSF-sponsored mathematics curricula were designed to reflect the best practices outlined in the NCTM *Standards.* Central to the creation of each of the programs was a focus on supporting and challenging all learners. The strategies here are not exhaustive, but offer ideas for helping students adapt to a different learning environment that in the end will better support their learning of mathematics. What is true across all the suggested accommodations and strategies is that they try to preserve the learning opportunity for students while providing the support that is needed. As described in *Principles and Standards* (NCTM, 2000), this is the Equity Principle: "Excellence in mathematics education requires equity—high expectations and strong support for all students" (p. 12).

PART III

PROFESSIONAL DEVELOPMENT AND TEACHER EDUCATION

This part of the book is organized around professional development in preparation to teach the NSF-funded middle grades mathematics materials. The first four chapters address curriculum-specific planning for professional development. These are followed by a chapter that looks at how teacher professional development needs change during implementation and another chapter that focuses on the professional development needs of administrators. Then four teachers talk about how their teaching changed as a result of these materials followed by a research mathematician's reflections on how his ideas about mathematics curricula and teacher preparation have evolved. The focus then shifts to how these materials can be used in preservice teacher preparation.

The following *Reflection Questions* relate to the four chapters on teacher growth and can be used for group discussion or for individual reflection.

TEACHER GROWTH QUESTIONS

1. What challenges did these teachers face as they started teaching a standards-based curriculum? Is there anything in common across the four stories in this regard? What did these teachers do to meet their particular challenge?
2. What seemed to motivate the teachers to make changes in their teaching? How did they know they were doing the "right thing?"
3. What changes did these teachers make in their classrooms? How were these changes related to the curriculum they were using?

4. None of these teachers seems to have been dragged "kicking and screaming" into using a standards-based curriculum. What advice do you think each would offer to a reluctant colleague?
5. What role did professional development play in the growth of each of these teachers?
6. What "outside" influences (either stated or implied) did each teacher encounter?
7. If you could ask these teachers questions, what else would you like to ask about their experiences teaching their curriculum?
8. How do these stories reflect your own teaching situation?

CHAPTER 14

AN OVERVIEW OF PROFESSIONAL DEVELOPMENT[1]

Elizabeth Phillips
Michigan State University, East Lansing, Michigan

Glenda Lappan
Michigan State University, East Lansing, Michigan

Yvonne Grant
Portland Public Schools, Portland, Michigan

Fran Arbaugh
University of Missouri, Columbia, Missouri

The implementation of new curricula in the classroom can serve as a powerful learning experience for teachers. For curriculum implementation to support professional development, plans must be designed that enable teachers to learn about, try, reflect on, and share information about teaching and learning in the context of implementing the curriculum with their colleagues.

—Loucks-Horsley, Love, Styles, Mundry, & Hewson, 2003, p. 127

In order for curriculum materials that were written to instantiate the National Council of Teachers of Mathematics *Curriculum and Evaluation Stan-*

A Decade of Middle-School Mathematics Curriculum Implementation, pages 141–150

dards for School Mathematics (NCTM, 1989; hereafter referred to as standards-based curriculum materials) to reach their full potential, teachers need to develop the knowledge and skill to use such materials to engage students in appropriate ways. Hence, the focus of professional development for teachers implementing a standards-based curriculum is to enable teachers to learn mathematics content and pedagogy needed to enact such curricula. A key aspect of supportive professional development (PD) is helping teachers learn to plan and enact instruction around these curricula. Several of the standards-based curricula are problem-centered. In such a problem-centered curriculum, contextualizing the mathematics is key—the mathematics is embedded in situations that may be real-world explorations, fantasy/story situations, or mathematical situations. Inquiry into the situations out of which the mathematics arises requires that teachers learn to orchestrate what we refer to as "inquiry-based instruction." A problem-centered curriculum is a radical departure from the traditional classroom in which most of us have learned or taught. Teachers will need time and assistance to become familiar with the mathematical understandings and skills that are embedded in the problems. They will also need to learn to use instructional strategies for helping students explicitly articulate these concepts and related skills.

Experience and research suggests that effective professional development models have some common characteristics. To be effective, professional development:

- Begins prior to curriculum implementation and is on-going
- Focuses on the standards-based mathematics curriculum that will be/has been adopted
- Develops teachers' knowledge of mathematics and pedagogy
- Models and reflects good mathematical pedagogy
- Addresses teacher concerns about change
- Involves teachers in reflecting and planning for improving their classroom practices
- Creates strong leadership
- Includes a plan for training new teachers as they join the district
- Reflects strong support from administration and parents
- Establishes ways in which teachers continue to interact around planning, teaching, and improving their mathematics program

Good professional development to support standards-based curricula weaves mathematics, pedagogy, and assessment together. As teachers continue to engage in professional development activities, attention shifts to student learning and the interrelationship between teaching and learning. Continued attention must be given to these concerns, including helping teachers to probe more deeply into the mathematics, their practice, and student learning. To make significant changes, professional development must provide opportunities for teachers to:

- Develop a deeper understanding of mathematics and a broader view of what constitutes mathematical knowledge
- Strengthen their knowledge of effective teaching strategies to promote student learning
- Explore classroom assessment aligned with inquiry-based instructional strategies

Professional development must be based on sound criteria and principles that have evolved from research and been verified in experience. The research discussed above as well as other research described by Loucks-Horsley et al. (2003), our own extensive experiences, and the *Professional Standards for Teaching Mathematics* (NCTM, 1991), serve as important references for our professional development design.

Three components—content knowledge, teaching and learning, and assessment—are core areas of an effective professional development model. Each comes to the foreground at appropriate times during professional development.

Mathematical Knowledge

Professional development associated with implementing standards-based mathematics curricula should include an emphasis on mathematical content. Standards-based materials and teaching practices create classrooms in which the unexpected happens. To steer the students' investigations and the summary of their explorations in mathematics terms, teachers need to be comfortable with the mathematics embedded in the problems. In addition, they need to build a more complete picture of how the mathematical ideas build on previous ideas and how those ideas in turn provide the foundation for the mathematics in later units and in subsequent grades. Good instructional decisions and practice rely on deep understanding of the mathematics that is embedded in the problems. To support teacher learning in professional development, effective instructional practices should be modeled as mathematics content is being learned.

Teaching and Learning

Teachers need to experience inquiry-based pedagogy in their professional development and reflect on the experience collectively and individually to identify and make explicit the components and actions that comprise such teaching. These experiences can serve as a model for their own teaching. They also need to be involved in sufficient conversations about teaching problem-centered materials to feel comfortable during implementation. More in-depth work on instruction after teachers have experienced teaching units for several months is needed. It is worth noting that when the focus shifts to pedagogy, teachers continue to develop their own understanding of the mathematics through conversations in which they analyze student work to assess levels of understanding. While the teacher and student books serve an important role in helping the teacher implement the curriculum within their classroom, teachers also need time away from their classrooms to talk with peers and to fully investigate the potential of the curriculum.

Assessment

Once teachers have begun using inquiry-based instruction, they recognize that traditional forms of assessment are insufficient to gauge student learning in this new environment. The standards-based curricula offer alternative forms of assessment to support teachers. Orchestrating different types of assessment requires new skills for teachers and therefore must be included in the professional development program. Figuring out how to assess and how to grade assessments is a concern that arises later in implementation. The appropriate time for assessment to be in the foreground in professional development is after the focus on teaching and learning. Placed at this point in the PD, the influence of assessment on teachers' reflection and the interaction between teaching and learning is enhanced.

CONTEXTS FOR PROFESSIONAL GROWTH

The activities (contexts) of a professional development program for teachers implementing a standards-based mathematics curriculum should have a strong emphasis on experiencing, planning, teaching, reflecting, and professionalism (Figure 14.1). These activities are typically designed to occur over several years and include summer and school-year activities.

Experiencing

To implement a problem-centered curriculum, teachers need deeper understandings of the key mathematical ideas and ways of reasoning that are embedded in solving the problems within a unit. Additionally, they need to

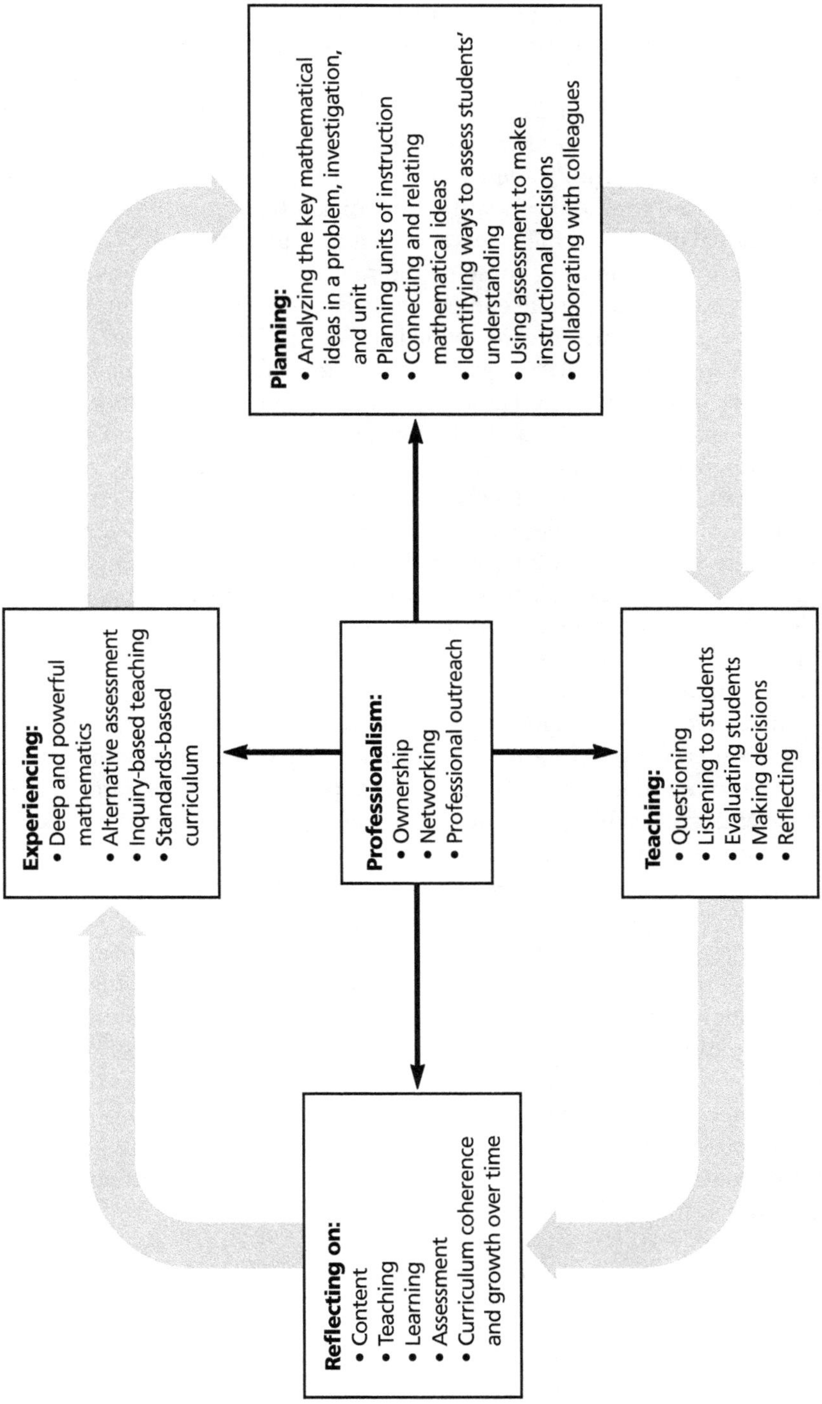

Figure 14.1 Contexts for professional growth.

see how understanding of these ideas develops over time and connects to content in other units. Thus during the workshops (or whenever teachers are being introduced to a new unit), teachers should experience the curriculum in a way that is similar to what their students will experience. This does not mean they need as much time for each problem, nor that they must do every problem. Problems for the workshops should be chosen to highlight the development of key mathematical ideas.

Professional development leaders should model good teaching. This allows teachers to focus on making sense of the mathematics needed to solve the problems posed. By setting the context with "How do you think your students might solve this problem?" the workshop leader can shift the focus to students' understanding. A powerful strategy for helping teachers with the mathematics and showing what students can do is to use examples of student work. This alleviates the anxiety of teachers who do not have confidence in their ability to do the mathematics. It allows teachers to ask questions they might be reluctant to ask. In such an environment teachers can and do learn the mathematics of a unit. Positioning the mathematics and the teaching through the lens of the student helps provide a comfortable environment for discussions on teaching and learning. It also helps teachers set higher expectations for what students can do and should be given opportunities to do.

Through follow-up discussions of the problems, the mathematical potential of the problems, the reasoning that students employ, and the connections that can be made become more explicit. Through such interactions teachers begin to value reflecting on such questions as, "What is the mathematics? At what stage are we in the development of understanding of the key idea?"

While effective teaching strategies are modeled, and occasionally discussed, when the unit is being presented (or taught) to the participants, it is often not until after the teachers have tried teaching the materials that they are ready for teaching practices to become their focus. Teachers will need help with developing a teaching model that supports student exploration with the goal of making sense of mathematical concepts and processes. Knowing how to engage students in a problem, how to assist and guide all students during their exploration of the problem, and how to help students make their understandings and their strategies explicit is very crucial to the development of the mathematics. Observing good teaching either live in a classroom or on video can provide a stimulus for discussions on teaching. Analyzing students' strategies can lead to conversations about how the classroom environment/discussion may have affected learning.

Planning

Planning is key for the successful implementation of a problem-centered curriculum. It is important to identify the mathematical concepts or strategies, their stages of development, and the time needed to develop these understandings. The power of a standards-based curriculum does not lie in any one activity or any one unit. Important ideas are studied in depth within a unit and further developed and used in subsequent units. It is both the depth of understanding within units and the careful building and connecting of the units that allow students to develop to their fullest mathematical potential. Initial planning can occur in the first summer, prior to the implementation of the curricula. However, teachers also need time during the year to plan, particularly with their colleagues. Planning sessions allow teachers to share problems they have experienced, learn new ideas from their colleagues, probe the mathematics more deeply, look for connections, and plan upcoming class sessions.

Teaching

The activity in which knowledge is developed and deployed....is an integral part of what is learned.

—Brown, Collins, and Duguid, 1989, p. 32

Teachers need to think critically about creating a classroom environment that fosters a community of learners. Posing problems that provide a challenge for the students, allowing students to explore the problem, and guiding class discussion on the solution of the problems requires the teacher to play many roles at the same time. Teachers need to ask effective questions that can probe students' understanding and at the same time they need to listen carefully to students. But equally important is getting students to ask questions of themselves and their classmates and to listen and respond to each other. Setting and achieving high expectations for understanding, problem solving, representing, and communicating for all students is a task that confronts teachers on a daily basis. Reflecting on one's practice with a lens on student understanding is important for teachers to make progress. At the end of each class period, teachers need to ask, "What evidence do I have that my students learned something? What did they learn?"

Reflecting

Professional development activities should model reflection. It is through reflection on their teaching and their students' understanding that teachers continue to grow in their capacity to build powerful mathematical experiences for all students. Planning with a colleague, peer coaching, observing a peer, or sharing with colleagues are some ways to encourage reflection.

Videotapes of lessons can serve as a catalyst for reflections. Caution must be exercised if videos are used. If teachers within a building agree to be video taped, then the focus should be students' learning rather than critique of the teacher. Centering conversations on student learning is a way to help a teacher think about his or her practice.

Finding the fine line between trying to help the students be successful with the problem and allowing the students freedom to explore a more open problem will take much reflection and planning time. Similarly, using a rich collection of student work can focus teachers' attention on the role and importance of the summary phase of a lesson. Many teachers, when they study a collection of student work, may say that all of the solutions are acceptable, or correct those that are not, and go on to the next lesson. But it is the analysis and comparison of the collection of student work that can bring the important mathematics to the forefront.

Student work can also be the way to center discussions and reflections on students' understandings. A variety of assessment tools should be used including mathematical reflections, quizzes, unit tests, projects, and district-wide instruments. To be effective, discussions on student learning should go hand-in-hand with teaching. Discussing student learning leads naturally to looking at the development of ideas over time. Talking and planning with colleagues in different grade levels provides the opportunity for teachers to build and share a coherent curriculum vision. Collaboration and reflection are key elements in the change process and in creating a community of teacher professionals that can support change over time.

Professionalism: Keeping the Momentum Going

Encouraging teachers to share their successes is one way for teachers to take ownership of the curriculum and their class. This can be done within the school by mentoring new teachers or through shared planning times with other teachers, through district newsletters, through on-line discussion groups, or by volunteering to speak at local or state meetings. Networking, going to professional meetings, and joining mathematics teachers' associations are all ways to continue to grow in mathematical knowledge and in pedagogical strategies. As teachers become comfortable with the curriculum, it becomes a natural part of the school fabric. But this is also the time when teachers and schools need to look more deeply at the curriculum and ask themselves "Can we do better?" This is the time when teachers should look deeper at the mathematics embedded in the problems and find ways to promote students' deeper understanding. Moving to this next level of implementation is a crucial step and one that is very often overlooked. Also overlooked is the teacher turnover that occurs in virtually all middle schools.

It is critical to develop a plan to provide professional development for new hires. It is equally as important, and not simple, to develop collaborative relationships with experienced mathematics teachers. Such relationships are mutually beneficial to new and experienced teachers.

It is through collaboration—building professionalism among teachers as a community—that progress is made and continued. Teacher leaders need to evolve to help maintain these communities. It is through these collaborative professional exchanges of sharing ideas, planning, examining student work, looking for gaps, and finding ways to make even bigger gains in student understanding, reasoning, and communication that we continue to move forward. In the early stages of implementation, the community may include the entire staff of mathematics teachers, but as implementation continues, it is likely that teachers will rely heavily on their grade-level colleagues for support, ideas, and guidance. Professional development must provide opportunities to ensure that these collaborations are able to continue throughout the implementation, even after the curriculum appears to be institutionalized at the school.

SUMMARY

During their careers, teachers play different roles—sometimes simultaneously. It is important that the professional development activities reflect the following four roles that teachers and teacher leaders play in creating more powerful mathematics classrooms. These four roles are:

- Teachers as students of mathematics
- Teachers as teachers of mathematics
- Teachers as collaborators with other teachers
- Teachers as facilitators working with administrators, parents, and other teachers

In a long-term professional development model, teachers will be in each of these roles. As schools adopt standards-based curricula, they need to make a specific plan for professional development of teachers. A summary of the major ideas and issues related to professional development and implementation of standards-based curricula discussed in this chapter is included in Figure 14.2. In addition, many of the standards-based curriculum projects have specific guidelines and suggestions for professional development related to using their curriculum, as will be discussed in the subsequent four chapters.

- Focus on mathematics content
  - Identify and understand the big mathematical idea(s) in a unit
  - Understand the development of mathematical concepts and reasoning from problem to problem, from investigation to investigation, and from unit to unit
  - Understand the development of algorithms
  - Identify the attention to skills embedded in the curriculum
  - Recognize connections between mathematical ideas within a unit, across units, across grades, and to applications
- Focus on inquiry-based instruction
  - Examine the instructional model recommended by the curriculum
  - Explore teacher questioning and listening techniques
  - View and discuss a standards-based classroom learning environment
  - Discuss appropriate pacing of the curricula
- Focus on assessing student learning
  - Examine various dimensions of assessment (content knowledge, mathematical disposition, work habits)
  - Examine various assessment components offered within the curriculum
- Address specific teacher concerns
  - Discuss and review various classroom management strategies
  - Examine strategies for grading, dealing with absences, second language students, special need students, etc.
- Support ongoing implementation
  - Establish time for teacher collaboration
  - Maintain community support through ongoing communication
  - Monitor student progress through systematic collection and review of data

Figure 14.2 Key issues related to professional development.

NOTE

1. This chapter is adapted from "Implementing Standards-Based Mathematics Curricula: Preparing The Community, The District, and Teachers" found at http://showmecenter.missouri.edu.

CHAPTER 15

KEY FEATURES OF *CONNECTED MATHEMATICS* FOR PROFESSIONAL DEVELOPMENT

Elizabeth Phillips and Glenda Lappan
Michigan State University, East Lansing, Michigan

The activity in which knowledge is developed and deployed . . . is not separate from or ancillary to learning and cognition. Nor is it neutral, rather, it is an integral part of what is learned. Situations might be said to co-produce knowledge through activity. Learning and cognition . . . are fundamentally situated.

—Brown, Collins, and Duguid (1989, p. 32)

Connected Mathematics (*CMP*) (Lappan, Fey, Fitzgerald, Friel, & Phillips, 1998, 2002, 2004, 2006) materials reflect this stance that teaching and learning are not distinct—"what to teach" and how "to teach it" are inextricably linked. The circumstances in which students learn affect what is learned. Thus the needs of both students and teachers are considered in the development of *CMP*. This curriculum is designed to help teachers, and those who work to support teachers, examine their expectations for students and analyze the extent to which classroom mathematics tasks and teaching practices align with their goals and expectations. The interactions between student and teacher materials comprise the first level of professional devel-

A Decade of Middle-School Mathematics Curriculum Implementation, pages 151–159

opment for teachers. We view the teacher materials as a conversation with teachers about possible ways to engage students in thinking and reasoning mathematically. The teacher materials are not a script, but are meant to peak teachers' interest and push their thinking about their instruction.

FEATURES OF CONNECTED MATHEMATICS PROJECT MATERIALS

Because the development of the mathematics in *CMP* is very different from traditional mathematics texts, a focus on the following three features of the *CMP* materials are essential in designing effective professional development for teachers:

- Important mathematical ideas are identified and studied in depth.
- The ideas are embedded in interesting contexts within an appropriate developmental sequence of problems, investigations, and unit.
- The *CMP* teaching model is developed to support teacher enactment of inquiry-based instruction.

Important Mathematical Ideas

The *CMP* curriculum consists of 24 units—eight units per grade—each developing a "key" idea or set of related ideas in depth. Each unit consists of a series of Investigations that are a sequenced set of Problems, followed by a set of homework problems called Applications–Connections–Extensions (ACE) and a Mathematical Reflection on the investigation. Each unit ends with a Looking Back, Looking Ahead *Reflection*. Once the important mathematical ideas have been identified for middle school, then for each grade level in *CMP*, a carefully selected subset of these important mathematical concepts, ideas, and related procedures are studied in depth. Students continue to deepen their facility with concepts and procedures by using them in new contexts in subsequent units and grade levels and, with each exposure, at a higher level of cognitive demand. This approach both deepens understanding of concepts and facility with procedures while connecting ideas to other important mathematics. However, it presents new challenges for teachers and a need for professional development that expands and affects both teacher knowledge and teaching practices.

Without a deeper understanding of concepts and how they are connected, students and teachers may view mathematics as many different techniques and algorithms to be memorized. For a student with such a view of mathematics, a problem-solving situation elicits a search for the correct algorithm. This approach contrasts sharply with a student who has developed the disposition to make sense of the situation, consider the nature and size of a solution, put together a solution path that makes sense, and examine the solution in light of the original question. Building such dispositions in students requires changing teaching practices as well. Taking time to allow the ideas studied to be more carefully developed means that when these ideas are met in future units, students have a solid foundation on which to build. Changing classroom practices to support inquiry and deeper understanding and skill is not easy. However, the payoff is that rather than being caught in an endless cycle of relearning the same ideas to some superficial, quickly forgotten level, students are able to connect new ideas to previously learned ideas and make substantive advances in depth of understanding and use of these ideas. The question becomes, "What teaching practices support students in realizing the full mathematical potential of *CMP*?"

A Problem-Centered Curriculum

Findings from cognitive science research support the theory that students can make sense of mathematics if the concepts and skills are embedded within a problem context. In *CMP*, mathematical ideas are embedded in a series of interesting problem contexts. As students explore the problems, they are asked not only to solve the problem, but also to defend their answers with cogent explanations, to look for patterns, to make generalizations, and to abstract from the experiences powerful mathematical ideas, problem-solving strategies, skills, and ways of thinking. The class together reflects on the investigation by critically examining, refining, and extending conjectures and strategies that have been derived from the investigation. The goal is for students to learn mathematics and learn *how* to learn mathematics. Most middle school teachers have not had much experience in processing such discussions. Professional development can help teachers build new images of and strategies for orchestrating classroom discussions that probe, challenge, and support the mathematical thinking of students.

A problem-centered curriculum not only helps students to make sense of the mathematics, it also helps them to process the mathematics in a retrievable way. Teachers of *CMP* report that students in succeeding grades remember and refer to a concept, technique, or problem-solving strategy by the name of the problem in which they encountered the ideas. For example, the Basketball Problem from the grade 7 unit, *What Do You Expect?* becomes a trigger for remembering the processes of finding expected values in multi-stage probability situations.

The CMP Teaching Model for Classroom Instruction Based on Inquiry

In a fully implemented problem-centered classroom, teaching practices change. Teachers challenge, guide, question and facilitate the learner in doing and making sense of the mathematics. This contrasts sharply with a more typical non-standards based role for the teacher as the source and demonstrator of knowledge. Clearly, changing practices this dramatically requires support from the curriculum and the teacher materials, as well as professional development opportunities to see, experience, discuss, experiment with, and assimilate such new practices. The teaching model we have developed, starts a teacher on his or her way to developing new ways of interacting with students and mathematics in the classroom.

When mathematical ideas are embedded in rich, contextualized problem-based investigations, the teacher has a critical responsibility for ensuring that students abstract and generalize the important mathematical concepts and procedures from the experiences with the problems. For more than three decades, we have been experimenting with ways to help teachers think about problem-centered teaching. As we developed the student materials and the supporting teacher materials, we took into account the demands of problem-centered teaching. Research and experimentation in classrooms led to the development and articulation of an instructional model that provides a lesson-planning template for teachers. This model looks at instruction in three phases: launching, exploring, and summarizing.

For each problem, the teacher *launches* the problem by providing a mathematical challenge to class, the class *explores* the problem, and then through examination of evidence, conjectures and arguments, consideration of alternatives, and generalizations of the mathematical results, the teacher and students work together to *summarize* the mathematics embedded in the problem. These three phases, Launch, Explore and Summarize (LES) are critical components of the *CMP* inquiry-based instruction. To carry out a successful lesson, careful planning, including a clear understanding of the mathematical goals for the lesson and strategies for each phase of the instruction, is needed.

Launch. In this phase, with a clear vision of the mathematical goals for the day, the teacher launches the problem with the whole class. This involves helping students understand the problem setting, the mathematical context, and the challenge. The following questions can help the teacher plan for the launch:

- What are students expected to do?
- What do the students need to know to understand the context of the story and the challenge of the problem?

- What difficulties may students encounter?
- What questions can facilitate student engagement and progress without giving away the problem solution?

It is critical that, while giving students a clear picture of what is expected, the teacher leaves the potential of the task intact. He or she must be careful to not tell too much and consequently lower the cognitive challenge of the task or turn it into an exercise, something routine. Another challenge for the teacher in the launch is to make certain she or he does not inadvertently cut off the rich array of strategies that may evolve from a more open launch of the problem. Giving an example of a possible approach is highly likely to do just that—stop creative thinking and focus the students on the teacher's strategy.

Explore. The nature of the problem suggests whether students work individually, in pairs, in small groups, or occasionally as a whole class to solve the problem during the *explore* phase. The Teacher's Guide makes suggestions for appropriate groupings. As students work, they gather data, share ideas, look for patterns, make conjectures, and develop problem-solving strategies. It is inevitable that students will exhibit variation in their progress. The teacher's role during this phase is to move about the classroom, observing individual performances and encouraging on-task behavior. The teacher helps students persevere in their work by asking appropriate questions and providing confirmation and redirection where needed. For students who are interested in and capable of deeper investigation, the teacher may provide extra questions or challenges related to the problem. We call such questions *Going Further* and we have provided examples in the explore discussion in the Teacher's Guide. Suggestions for helping students who may be struggling and those who have differentiated learning needs are also provided in the Teacher's Guide. The explore part of the instruction is an appropriate place to attend to special needs. The following questions can help the teacher prepare for the explore phase:

- What grouping of the students to explore this problem is appropriate and will be productive? (Individuals? Pairs? Groups? Whole class?)
- What materials will students need?
- How should students record and report their work?
- What different solution strategies might they use?
- What questions are likely to encourage student conversation, thinking, and learning?
- What questions are likely to focus students' thinking if they become frustrated or off-task?
- What questions can challenge students who think they have the initial question "answered"?

During the explore phase, as the teacher moves about the classroom observing, questioning, and collecting data for the summary, the following questions need attention:

- What difficulties are students having?
- How can I help without giving away the solution?
- What strategies are students using? Are they correct?
- How will I use these strategies during the summary?
- Which student in a group should I call on to present the group's solution and problem-solving strategies?

During the explore phase teachers are assisting students in their learning, assessing students' progress, and collecting information to guide the summary.

Summarize. It is during the summary that the teacher guides the students to examine their mathematical findings from the problem (including the mathematical tools, ways of thinking, explanations, conjectures, arguments, examples, and generalizations they used or made in the process of solving the problem) and to connect their new understanding to prior mathematical experiences and problems in the unit. The *summarize* phase of instruction begins when most students have gathered sufficient data or made sufficient progress toward solving the problem to support a good mathematical discussion. In this phase, students present and discuss their solutions as well as the strategies they used to approach the problem, organize the data, and find the solution. During the discussion, the teacher helps students enhance their conceptual understanding of the mathematics in the problem and guides them in refining their strategies into efficient, effective, generalizable, problem-solving techniques or algorithms.

Although the summary discussion is guided by the teacher, students play a significant role. Ideally, they pose conjectures, question each other, offer alternatives, provide reasons, refine their strategies and conjectures, and make connections. The goal is for students, as a result of the discussion, to become more skillful at using the ideas and techniques that come out of the experience with the problem.

At appropriate places in the development of an idea, the summary can end by posing a problem or two that checks students' understanding of the mathematical goal(s) that have been developed to this point in time. *Check For Understanding* questions occur occasionally in the summary in the Teacher's Guide. These questions help the teacher to assess the degree to which students are developing their mathematical knowledge. The following questions can help the teacher prepare for the summary:

- How can I help the students make sense of and appreciate the variety of methods that may be used?

- How can I orchestrate the discussion so that students summarize their thinking about the problem?
- What concepts or strategies need to be emphasized?
- What ideas do not need closure at this time?
- What definitions or strategies do we need to generalize?
- What questions can guide a discussion and examination of the generalizability and efficiency of the strategies used by students without closing out students who need to continue to work for a while longer with more intuitive strategies?
- What connections and extensions can be made?
- What new questions might arise and how do I handle them?
- What can I do to follow up, practice, or apply the ideas after the summary?

Mathematical Reflections

It is also important for learners to reflect at critical points in their development on what they have learned. Therefore, at the end of each investigation a *Mathematical Reflection* is provided for the teacher and students. This reflection is a set of questions designed to help students reflect on what they have learned, why it is important, when it is useful, and how it fits with prior knowledge. The questions in the reflections point to the "big ideas" and how they fit together (abstracting and connecting); to skills (how-to); to decisions to be made and how one makes them (when-to); to issues about how these problems are similar to, and different from, problems encountered earlier (connecting, discriminating, and elaborating); and finally, to questions to ask in similar situations. The reflection questions and answers are first discussed by the class and then the students write their individual answers to these questions in their notebooks as part of their self-assessment. Each unit contains three to five investigations and hence three to five Mathematical Reflections. The unit ends with a reflection called, *Looking Back, Looking Ahead* (Lappan, Fey, Fitzgerald, Friel, & Phillips, 1998). This final reflection pulls the major ideas together, makes their connections transparent, and organizes the mathematics in a connected set of ideas, concepts, and procedures. Teachers in general find problem-solving, inquiry oriented materials very helpful to their own understanding of the mathematics they teach. The professional development challenge is to help teachers build powerful new classroom routines and ways of thinking that support students in their learning from such materials.

MATERIALS FOR SUPPORTING TEACHERS' IMPLEMENTATION OF CMP

From the beginning, we have viewed *CMP* as a curriculum for both teachers and students. The extensive field-testing of *CMP* has helped produce teacher materials that are rich with examples and suggestions from teachers' successful strategies, classroom dialogues and questions, potentially difficult points or tricky transitions, and examples of student solutions. We have used these experiences to develop an *Implementing and Teaching Guide* as well as a teacher guide for each unit as a part of the professional development support.

Implementing and Teaching CMP Book

This book provides a detailed introduction and overview of the program including the philosophy, mathematical and problem-solving goals, and instructional model for the curriculum. It also provides specific recommendations for classroom management, addressing special needs students, communicating with parents, and professional development resources.

Unit Teacher Guides and Assessment Resources

Each Teacher Guide engages teachers in a conversation about what is possible in the classroom around a particular lesson. Suggestions are made about how to engage the students in the mathematics task during the launch of the problem, how to promote student thinking and reasoning during the exploration of the problem, and how to work with the students to summarize the important mathematics embedded in the problem. Support for the Launch–Explore–Summarize instructional and learning sequence occurs for each problem in the *CMP* curriculum. Actual classroom scenarios are occasionally included to help stimulate teachers' imagination and vision about what is possible.

Each Teacher Guide also contains, in a section called the Mathematics of the Unit, an extensive discussion of the mathematics underlying the Investigations. References are made to this mathematics essay throughout the Teacher Guide to give teachers mathematical support at the point in which it is needed. The Teacher Guide also articulates the mathematical and problem-solving goals for each investigation and problem, gives connections to other units, includes an extensive and diverse assessment package, lab sheets, and other needed resources, as well as in-depth teaching notes. Homework suggestions are given and coded as to whether tasks are essential, connections, or extensions.

Since these documents contain important information about teaching and learning, we envision that teachers will refer to these documents frequently throughout their teaching of *CMP.* But for many teachers, written materials, while necessary, are not enough. Our experience suggests that teachers also need to engage in professional development (PD) with other teachers—particularly PD experiences that promote teacher learning through collaboration; these types of PD experiences are described in chapters 14 and 19 of this book.

In our work with *CMP* teachers, we have found it particularly helpful if teachers can observe each other teaching the same lesson or if one or two teachers would volunteer to be video taped. These video-taped lessons become part of the analyzing, reflecting and planning stage of instruction. It is also useful for teachers to grade and analyze student work for the same problem or assessment piece. Student's mathematical reflections give an additional way of analyzing student sense making.

CONCLUSION

While we have tried to provide extensive help for teachers in our curriculum materials, this is not sufficient. Curricula, like *CMP*, require change in the ways teachers think about teaching, learning, and assessment. It is important that districts provide long-term professional development as recommended in other chapters in this book. (See chapters 5, 8, 14, and 20.) For those schools that implement *CMP* accompanied by long-term professional development there is growing evidence[1] that students' mathematical learning increases.

Finally, encouraging teachers to share struggles and successes is one way for schools and districts to promote teacher ownership of the *CMP* curriculum. Sharing can be done within the school through mentoring new teachers, shared planning times with other teachers, district newsletters, or online discussion groups. Networking, going to professional meetings, workshops, special courses, and joining mathematics teachers' associations are all ways to continue to grow in mathematical and pedagogical knowledge and strategies.

NOTE

1. See 2006 Research and Evaluation Data for *CMP*, Prentice Hall Publishers

CHAPTER 16

KEY FEATURES OF *MATHEMATICS IN CONTEXT* FOR PROFESSIONAL DEVELOPMENT

David C. Webb
University of Colorado at Boulder, Boulder, Colorado

When implementing any new curriculum, many aspects of the program are new to teachers: the mathematical goals of activities, the connections between activities, the relative importance of questions and tasks, the location of instructional support, and how assessments should be used, to name just a few. The purpose of this chapter is to discuss the fundamental principles of Realistic Mathematics Education articulated throughout *Mathematics in Context* (National Center for Research in Mathematical Sciences Education & Freudenthal Institute, 1997, 1998, 2006), to support the selection of appropriate professional development activities and promote greater fidelity in teacher use of the curriculum.

Mathematics in Context (*MiC*) was designed to exemplify the theory of Realistic Mathematics Education, an approach to teaching and learning mathematics originally proposed by the Dutch mathematician Hans Freudenthal and articulated through several decades of educational research.

A Decade of Middle-School Mathematics Curriculum Implementation, pages 161–171

Professional development activities specific to *MiC* should reflect the main points of emphasis that distinguish Realistic Mathematics Education and the design of *MiC* (Webb & Meyer, 2007).

From the perspective of those who have been involved in the design of MiC and have supported its implementation in schools, teachers' professional development experiences should address: (a) the function of problem contexts; (b) the rationale for various mathematical models and tools; (c) the sequencing of activities to promote progressive formalization; and (d) the multi-dimensional approach to assessing student understanding. These principles are discussed in greater detail throughout the rest of this chapter.

THE USE OF PROBLEM CONTEXTS

A cursory review of a *MiC* student book reveals its most noticeable feature, that mathematical activities are grounded in "realistic" contexts. Such contexts involve everyday life settings or fictitious scenarios, although mathematics itself can also serve as a context of interest. What is not as discernible flipping through the materials is the way in which these problem contexts serve different purposes in *MiC*. Problem contexts have been used in many mathematics curricula to convey the relevance of mathematics, to motivate students to either propose new solution strategies or apply known skills to solve problems. In *MiC*, however, a problem context can be used to suggest a specific mathematical model that can be used across many other mathematically similar situations (Meyer, Dekker & Querelle, 2001). Some of the more well-known examples of this approach are: the sharing of submarine sandwiches to suggest rectangular fraction bars (see Figure 16.1), using download bars for computer software to suggest the use of percent bars, keeping track of customers' menu orders to suggest "notebook notation" (see Figure 16.2), and tracing the paths of jumping frogs to suggest a visual representation that supports solving equivalent algebraic expressions. These problem contexts, therefore, are not interchangeable with other seemingly related contexts or methods of presentation. For example, if one were to replace the "fair sharing" of sub sandwiches with pizzas, the mathematical model that would emerge would be a circular representation instead of the fraction bar, which has connections to the double number line and the percent bar.

It is also important to recognize that the problem contexts used in *MiC* are "realistic" and not necessarily "real world." Problem contexts are selected and adapted to motivate students' own strategies, their use of visual representations, and suggest related mathematical connections. Rather than convey the complexities of the real world, problem contexts in *MiC* are in-

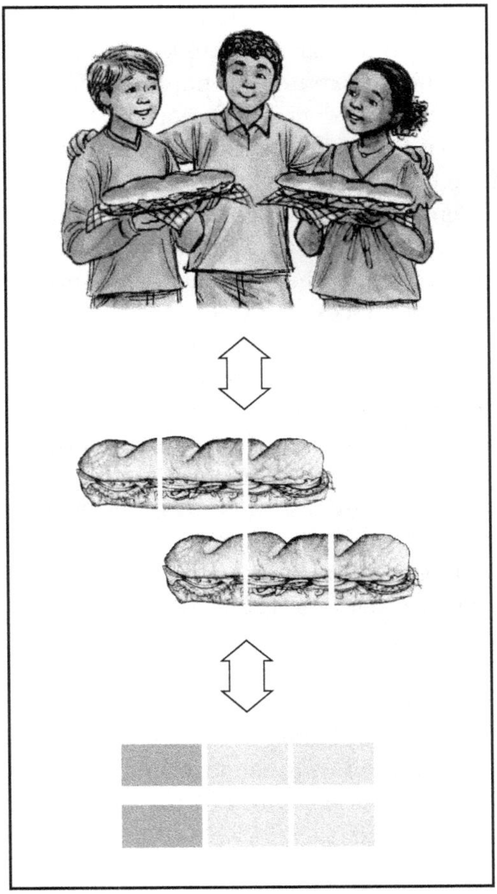

Figure 16.1 From sub sandwiches to fraction bars (adapted from Van Galen et al., 2006, p. 2).

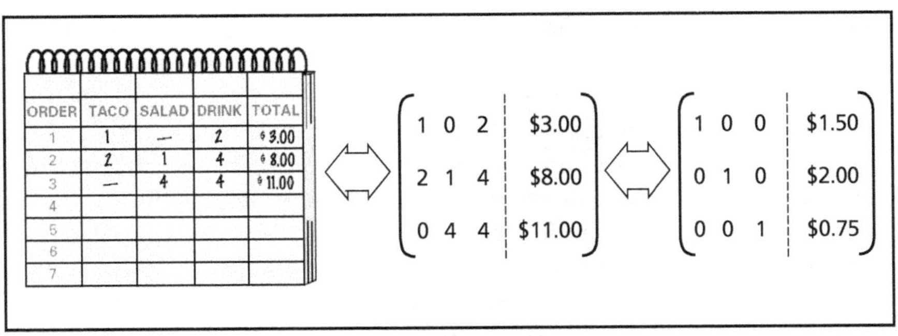

Figure 16.2 From notebook notation to Gaussian row operations (adapted from Kindt, Abels, Dekker, et al., 2006, p. 27).

tended to make the situation and related mathematics real in the students' mind. The contexts help to make the models and tools memorable and a natural part of the students' strategic repertoire. Waiters rarely, if ever, use notebook notation to take orders. It is important, therefore, to understand that many of the situations in *MiC* are idealized to motivate student reinvention of specific mathematical models that contribute to the development of more formal mathematics. This approach of connecting contexts to mathematical models gives relevance and meaning to the formal mathematical representations and strategies that students will eventually encounter.

Given teachers' experiences with word problems in conventional mathematics curricula, it is easy for teachers to see problem contexts only as an occasion for students to apply their procedural knowledge of mathematics (e.g., think of how coin and mixture problems have been used to apply knowledge of solving systems of equations). In contrast, many problem contexts used in *MiC* are far more than an opportunity for application; indeed, they are opportunities for students to derive new mathematical strategies and tools to solve other problems. From the teachers' point of view, since these strategy-building contexts have a different purpose they must be supported with instruction that emphasizes the essential features of the tool being developed. In these cases, the development of the model is the primary goal and the introductory context is secondary. Eventually, students will have an opportunity to apply the model in other problem contexts to appreciate both its power and limitation. Rather than having students "discover" these models on their own, the teacher has an instrumental role in supporting students' *guided reinvention* of the models through the use of questions, examples, and counterexamples that can clarify how the model could be used.

THE USE OF MODELS

The introduction of and emphasis on mathematical models in *MiC* requires that teachers attend to a greater range of strategies that are developed and used by students. From the teachers' perspective, the teaching of common algorithms might seem safer and more desirable due to the predictability of established procedures and the narrow window of potential errors that might result (e.g., think of the cross-multiplication algorithm for solving proportions). When using *MiC*, teachers will invariably encounter a tension between teaching for student recall of common algorithms and teaching for student understanding of formal mathematics through the guided reinvention of mathematical models. Although these models are more likely to promote student sense making and understanding, their development requires a more careful, student-centered pace of instruction.

When students use models such as the double number line, percent bar, or ratio table to solve problems (see Figure 16.3), the models are more apt to facilitate strategies that are at the students' comfort level (e.g., doubling, halving, use of decade numbers, and so on): models often can be used in many ways to achieve a valid result. So from the students' perspective, models offer choice and ownership in the solution strategy. Number models such as the double number line and percent bar also have the additional benefit of visualization of number and development of student number sense. These additional benefits for students, however, require an increased awareness from the teacher to interpret a wider range of student responses. Teachers should be given ample opportunities, therefore, to discuss the instructional dilemmas that arise with multiple representations and student-centered instruction (Meyer, 1999).

It is also important for teachers to explore, for themselves and with students, the interrelationships among the mathematical models used in *MiC*. For example, the double number line, percent bar, and ratio table all involve either pairs of numbers or ratios. In the example below, the double number line pairs two units, the number of blocks and the number of miles assuming that 1 block is equivalent to one-eighth of a mile. With the percent bar, the amount of time it takes to download a computer file is paired with the percent of the file that has been downloaded. In this case, the visual model of the percent bar is consistent with the download bar that students have encountered (or will eventually encounter) when installing software or downloading files from the Internet. The ratio table is used to organize equivalent ratios; in the case shown here, one package contains 15 pencils. Each of these number models organizes pairs of numbers so that students can explore equivalent ratios and develop their proportional reasoning.

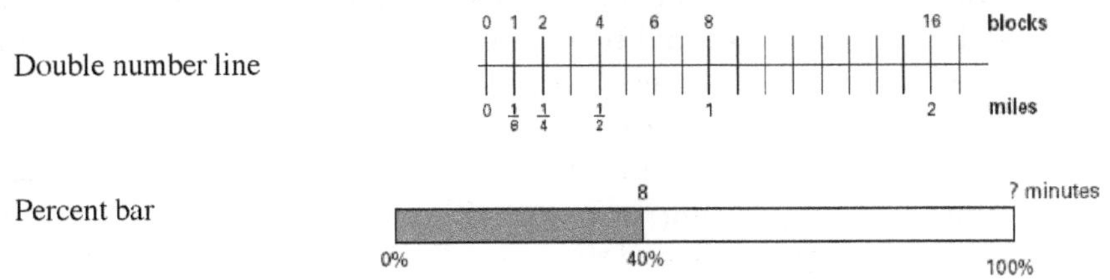

Ratio table

Packages	1	2	4	8	9	90
Pencils	15	30	60	120	135	1,350

Figure 16.3 Number models in *MiC* (adapted from Kindt, Dekker, Burrill, & Romberg, 2006).

However, each of these models has distinctive features that teachers need to explore with students. For example, the placement of ratios on the double number line and percent bar must be ordered, whereas with the ratio table the ratios can appear in any order, which promotes student number sense and strategic competence. Teachers (and students) should also be aware of the limitations of various models. An area model is useful for organizing the multiplication of mixed numbers and binomials, but when the binomials involve subtraction, the model is not as helpful. Even though these various models support student understanding, they should be examined with the same critical eye we might expect students to use when trying to make sense of common algorithms.

Models such as percent bars, ratio tables, and the like constitute new mathematical terrain for most teachers. To appreciate the instructional value of these models and their limitations, teachers have to learn, as students, how to solve problems using such representations. Teachers who understand the interrelationships and inherent limitations are better prepared to select and design new problems that can support student understanding. Eventually, students should see the need to acquire more formal strategies to solve problems. For example, the ratio table is easier to use with whole numbers and friendly ratios, but teachers can selectively introduce problems involving more difficult number pairs to motivate students to explore more formal strategies (e.g., scale factors, solving proportions, and so on). Nevertheless, while students will eventually develop a preference for more formal approaches they should still be able to use models when appropriate. Building awareness of the uses and limitations of models will enable teachers to be better able to promote students' movement to more formal strategies. It is also worth noting that these models are "thinking tools," to support student problem solving, and should not be seen as something for students to master. Since these models support student number sense they are often completed using scrap paper or mental operations. Likewise, the models students use should be regarded as strategies—models are not "answers" that must be completed with exacting precision (e.g., when students use a percent bar the location of 25%, 40%, or 50% should be estimates, not exact locations found with a ruler). Furthermore, since models are student-centered thinking tools they will represent a student's comfort level with numbers, patterns, and strategies. Each student will choose to use different models in different ways.

PROGRESSIVE FORMALIZATION

As models are applied to more mathematically sophisticated problem contexts, a model changes in character and becomes an entity in itself and is eventually used as a basis for mathematical reasoning on a formal level. As articulated in Realistic Mathematics Education this is the design principle of *progressive formalization.*

The "process of going from the concrete to the abstract" has been espoused in education psychology literature for many decades (cf. Piaget, 1952). Conventional middle grades mathematics curricula, however, have generally portrayed learning sequences in terms of related abstract representations to develop procedures and algorithms rather than giving students opportunities to use concrete models to give meaning to mathematical abstractions. In contrast, *MiC* instructional sequences are conceived as "learning lines" in which starting points are used to elicit necessary informal, often concrete, representations to support student sense-making of important mathematical abstractions (e.g., using sub sandwiches to suggest the mathematical model of fraction bars to represent fractions). When appropriate, students are introduced to strategies and models that are progressively more formal. Students are also encouraged to refer back to less formal representations as needed to deepen their understanding of the abstract-symbolic.

Teachers need to experience how sequences of problems from *MiC* embody progressive formalization. The informal models evoke class discussions that are accessible to all students, and connect familiar ideas from earlier experiences (e.g., sharing, division, equal parts) with pre-formal representations and tools (e.g., fraction bars; comparing, adding and subtracting fractions; double number line; and ratio table). When these relationships between problem contexts and related tasks requiring a student response are illustrated to teachers, they are better equipped to support *MiC* instructional sequences with additional probing questions and feedback that connect student knowledge and promote student understanding. More importantly, teachers must also realize that in *MiC* progressive formalization is often accomplished across several units within and across grade levels.

Since teachers using sixth-grade *MiC* units will, in many cases, be introducing models at an informal/preformal level they should be cognizant that the development of formal mathematics occurs later in the year in other units or in subsequent years. To promote and achieve student understanding, students should be given time to progress in formalization at their own pace and not be forced to use formal mathematics before they are ready to do so. Even though this "delayed formalization" can be a frustrating experience for teachers who are typically more familiar with the formal mathematics, the instructional design principle of progressive formalization does not imply a race toward formal mathematics. As suggested by Liping Ma's (1999) research, many teachers who are familiar with formal mathematics procedures, such as division of fractions, are challenged when they are prompted to present a conceptual model of the procedure to students. Formal mathematics is indeed important, but it is also essential for teachers to know the informal and preformal representations that can lead students to understanding of formal mathematics. It is essential, therefore, to give teachers ample opportunities to experience how sequences of activities and tasks in *MiC* are used to promote connections between students' informal knowledge and formal mathematics, leading to greater student understanding.

PROMOTING AND ASSESSING MATHEMATICAL REASONING

To teach for student understanding, in addition to valuing informal and preformal models, teachers also need to realize that students necessarily must encounter tasks that require more than recall knowledge. Students need to demonstrate that they can apply their knowledge in new ways. In *MiC*, the tasks posed during both instruction and on assessments are designed to elicit three different types of student reasoning:

Level 1: Recall of procedures, concepts, and definitions. These problems elicit recall knowledge of facts, mathematical properties, routine procedures, and standard algorithms. In general, Level I problems assess basic knowledge and procedures that may have been emphasized during instruction. On a written unit assessment, a Level I problem may use a context from the unit and closely resemble other problems that the student has already encountered.

Level 2: Making connections. At this level, students start making connections within and between the different domains in mathematics. This problem type is often open to various solution strategies to promote student choice of mathematical models. At this reasoning level, students are also expected to choose among different representations that are suitable for particular situations and need to be able to interpret various mathematical statements.

Level 3: Mathematization, generalization, and insight. Questions at this level require students to mathematize situations: that is, recognize and extract the mathematics embedded in the situation and use mathematics to solve the problem. In some cases, students will need to develop their own models and strategies. In other cases, students might need to construct mathematical arguments, generalizations, and proofs. Level III problems are always open-ended and typically more than one answer is possible, depending on how students justify their response.

The assessment pyramid shown in Figure 16.4 is a model used to suggest how assessment tasks that elicit these various levels of student reasoning should be distributed. The assessment pyramid is also used to highlight

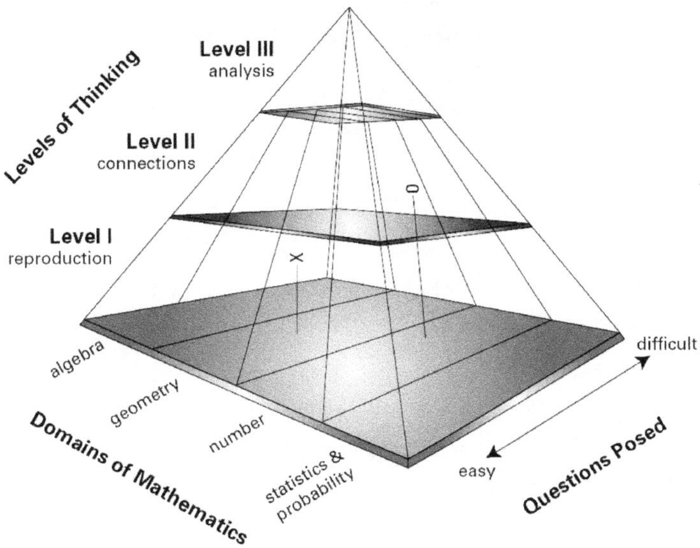

Figure 16.4 The assessment pyramid (Dekker, 1997).

three different dimensions that can be used to classify problems: mathematical content, level of reasoning, and difficulty level. As shown in the pyramid, Level I tasks are not necessarily easy. For example, Level I tasks may involve recall of procedural knowledge, but complicated by the choice of numbers or symbols. In *MiC*, tasks that are used to assess student understanding should ideally fill the pyramid. The shape of the pyramid is meant to suggest the relative distribution of assessment tasks for each level according to the relative volume for each level. That is, over 50% of the assessment tasks should be level I, 35% should be level II, and 15% should be level III.

Teachers who are unaware of these different types of reasoning objectives are likely to pass over higher-level reasoning tasks used during instruction or on unit assessments. Although teachers are not necessarily adverse to challenging students during instruction with questions that elicit more sophisticated reasoning, teachers are reluctant to pose similar tasks to students on written assessments. In fact, teachers in their first few years of implementing *MiC* often note that the assessments expect students to "take the mathematics to a new level," and are caught off-guard with higher-level tasks that require students to demonstrate that they could apply their mathematical knowledge in new problem contexts.

To increase teacher awareness of the different levels of reasoning elicited by problems throughout *MiC*, every teacher guide includes a Goals and Assessment table (found on pages xvi–xvii of every teacher guide) aligned with that unit. In addition, throughout each teacher guide, assessment

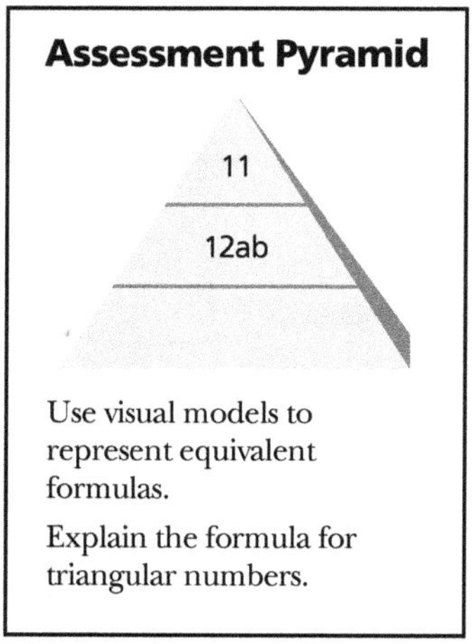

Figure 16.5 *MiC* assessment icon (adapted from Webb, Krusi, Palha, & Dekker, 2006).

pyramid icons (see Figure 16.5) highlight opportunities for ongoing assessment (i.e., instructionally embedded assessment). The three levels on the icon mirror the three levels of reasoning discussed above. In the example shown in Figure 16.5, problems 12a and 12b assess Level II reasoning, and problem 11 assesses Level III reasoning.

Building student confidence in problem solving comes through ongoing engagement with new problem contexts in which students can either propose their own solution strategy or draw from previously learned content to mathematize the problem. It should be expected, however, that students initially will be challenged, and perhaps frustrated, with problems that require them to apply their knowledge in new ways—an experience that is distinctively different both cognitively and emotionally than responding to questions that require only recall of facts or procedural knowledge. Teachers who are aware of the different reasoning levels of questions that appear throughout *MiC* units are able to support students as they encounter new challenges. With such awareness, teachers are better prepared to shift to an instructional stance that encourages student persistence in problem solving (Her & Webb, 2004). If the degree of student frustration is exceptionally high, teachers can suggest connections between the problem under consideration and mathematics topics they have previously discussed.

SUMMARY

Although the middle grades mathematics curricula funded by NSF were designed to support similar instructional objectives, each program's development team and their collective perspective on teaching and learning mathematics influenced the style, structure, and format of the resulting curriculum. Given that *MiC* was developed to reflect the principles of Realistic Mathematics Education, distinctive differences exist between *MiC* and the other curricula.

The purpose of this chapter was to highlight *MiC*-specific design principles that should be made explicit when designing professional development experiences for teachers—the role of problem contexts; the use of mathematical models, how activities are sequenced to promote progressive formalization, the multi-dimensional approach to promoting and assessing student understanding. Even though each of these design principles is articulated in the *MiC* Teacher Implementation Guide (Meyer, 2006) and the teacher guide for each unit, teachers need opportunities to explore and discuss how these design principles are reflected in the structure of *MiC* units and in the program as a whole. To understand how problem contexts can be used to introduce and develop powerful mathematical models, teachers need opportunities to explore how problem contexts are used in different ways in *MiC*. To appreciate the value of the mathematical models introduced and used in *MiC*, teachers need opportunities to use the models to solve problems and discuss how such models increase student sense making and access to mathematics. The collective experience and insight from those who have supported teacher implementation of *MiC* through workshops, user conferences, and school-based support confirm that these design principles are effective themes for organizing professional development experiences, selecting appropriate activities from *MiC* units, and guiding discussions of pedagogy and assessment.

CHAPTER 17

KEY FEATURES OF *MATHSCAPE* FOR PROFESSIONAL DEVELOPMENT

Fred Gross and Emily R. Fagan
Educational Development Center, Newton, Massachusetts

Creating a professional development plan is central to preparing for and sustaining the successful implementation of new curriculum materials. A sound professional development plan for implementing *MathScape: Seeing and Thinking Mathematically* (Education Development Center, Inc., 1998, 2005) will incorporate effective techniques such as discussions and hands-on activities that uncover the key philosophical principles and the overall structure of the curriculum. This chapter outlines key characteristics of *MathScape* that distinguish it from similar programs, and that should guide the design of any professional development that prepares teachers to use *MathScape* in their classrooms.

MathScape's developers were guided by a vision of mathematics classrooms in which *all* students are successful in learning significant mathematics. This vision encompasses many dimensions, best expressed in the words of Kleiman, Fagan, Janssen, Brodesky, and Tobin (2007):

A Decade of Middle-School Mathematics Curriculum Implementation, pages 173–182

- "students building their knowledge of essential mathematics concepts, skills, and facts so they could apply this knowledge with fluency, flexibility, and deep understanding;
- "teachers taking on multi-faceted roles by actively engaging students in rich mathematical investigations, facilitating discussions in which students communicate and clarify their mathematical ideas, and assessing students' understandings;
- "meeting the developmental needs of middle school students by sequencing instruction to move from concrete to abstract, and by helping students transfer from their own informal vocabulary to the more precise and powerful language of mathematics;
- "engaging and motivating students through contextualized lessons, through creative and collaborative work, and through opportunities to share their own knowledge and experiences" (p. 98).

Professional development for *MathScape* should help teachers understand this vision, see how it is embedded in *MathScape's* activities, lessons and units, and help them bring it to life in their classrooms. In order to achieve this, developers and others who have helped districts and schools implement *MathScape* recommend emphasizing four key areas:

- Context to Support Learning: How does *context* serve to engage middle school students in mathematics, and help them build understanding across content areas?
- Communication about Mathematics: What is the rationale for structuring *discourse*, oral and written, as a key feature of every lesson?
- Design of Lessons and Units: How are the lessons and units structured? How do they advance learning and address the needs of a range of learners?
- Assessment: How does *MathScape* incorporate assessment? What is the philosophy behind the assessments, and what are the components of informal and formal assessment?

CONTEXT TO SUPPORT LEARNING

In developing *MathScape*, the authors asked themselves, "What interests and engages students? How can we incorporate this into the materials?" They believed that middle school students want to see a purpose in what they are learning, and figure out how the world works and their place in it—and that these beliefs should help frame the approach to learning and studying mathematics. Therefore, *MathScape* sets rigorous mathematics within accessible and interesting contexts. A situational context provides a framework for students to study mathematical concepts and skills and apply them strategically in various situations. It helps them see the value of mathematics in addressing real needs. Students make decisions based on the mathematics within the particular context. And, by studying the concepts set in various contexts, students are able to make connections that deepen their understanding and foster flexibility in their mathematical thinking. For example the grade 7 unit *Buyer Beware* (Bousquet, Janssen, & Brodesky, 2005; see Figure 17.1) uses the context of consumer awareness and advocacy to study rate, ratio and proportion. Throughout the unit, students examine consumer issues such as advertising, comparative shopping, and event planning. Through these investigations, students come to understand that proportional relationships are multiplicative rather than additive. In developing these proportional reasoning understandings, students are prepared to apply them to different contexts in subsequent units.

Compare Ratios in Trail Mix

Dwight Middle School is planning an Autumn Festival. The seventh grade is planning to buy trail mix to sell at the festival. They want to decide if one of two brands, Mountain Trail Mix or Hiker's Trail Mix, contains more chocolate by weight or if both contain the same amount. The *Buyer Beware* research group has collected the information shown on the two brands. Note: The mixes contain only dried fruit and chocolate chips.

1. What is the ratio of chocolate chips to dried fruit by weight in Mountain Trail Mix?
2. What is the ratio of chocolate chips to dried fruit by weight in Hiker's Trail Mix?
3. Are the ratios of chocolate chips to dried fruit for each type of mix the same or different?
4. Does one trail mix brand contain more chocolate per ounce than the other? If so, which one? How do you know?

Figure 17.1 A context to support study of rate, ratio and proportion (Bousquet, Janssen, & Brodesky, 2005).

In other units, students expand and apply proportional reasoning (ratio and scale) and geometric measurement skills (area, perimeter and surface area) as they design houses to cost and design specifications. They also build understanding of related concepts including scale, similarity, triangle relationships, and trigonometry by exploring problems in designing staircases, roadways, and ramps. This variety and range of situations ensures that the students' skills and understanding progress, and allows students to make connections between concepts and across contexts to the bigger mathematical idea of proportional reasoning. Through such experiences, students begin to see the importance of the relationship between the mathematics and the situation, and apply their mathematical thinking. What happens, for example, to the original situation if a change is made to a number, or an angle, or a variable?

Providing contexts for learning mathematics may feel foreign to teachers accustomed to conducting lessons that focus on developing procedures or algorithms. Yet it is important to realize that the contexts allow students to *make sense* of the mathematics and feel more comfortable with it. Building awareness of the value of context better prepares teachers to use it effectively in math instruction.

In some *MathScape* units context is not as central to the unit, and students build understanding through a slightly different framework. In *From Wholes to Part* (Karnowski, Fagan, Thorpe Breeze, & Marrero, 2005) and *Beside the Point* (Karnowski, Benson, Janssen, Doherty, Fagan, & Thorpe Breeze, 2005), two units about fractions and decimals, the framework is not one particular situation, but rather how number line models and area models can help to build conceptual understanding of rational numbers. In such units, the context might vary and the consistent thread is applying these models to a variety of cases.

COMMUNICATION ABOUT MATHEMATICS

The National Council of Teachers of Mathematics, in *Principles and Standards for School Mathematics* (2000), articulates the importance of communication in the study of mathematics in promoting the recognition of connections among ideas and the reorganization of knowledge. The *MathScape* curriculum recognizes the importance of routinely sharing ideas and perspectives. Mathematical literacy requires the ability not only to know mathematical ideas and concepts, but to communicate that understanding to others. *MathScape* provides opportunities to build communication skills in three important ways: small group investigation and discussion, whole-class discussion facilitated by the teacher; and written ideas, justifications, and reflections on mathematics problems. Through classroom discourse, written and oral, students can identify, justify, and challenge their thinking, a process that leads to better understanding of the complexities of the mathematics.

Oral Discourse

In the 7th-grade unit *Making Mathematical Arguments*, students are asked to develop mathematically sound arguments about integer operations. For example, they explore the question: Is it true that when adding a positive integer and a negative integer together the result will always be negative? Students present arguments and counterarguments to support this statement as true or false. The rich discussion among students in small groups and in the whole group draws out their ideas and mathematical understandings about integers and provides a forum to share and refine their mathematical arguments. In this and many other instances, students examine and revise their statements based on the thinking of others. Sharing ideas through discourse provides a way to see and explore the flexible nature of mathematics and the importance of developing consistent rules.

Written Discourse

Writing in mathematics provides students with other opportunities to interpret, express, and share mathematical ideas. The *MathScape* curriculum emphasizes hypothesizing mathematical ideas and articulating student-generated justifications for ways of thinking about the content. Written demands take various forms, such as descriptive writing with short explanations, and expository writing in the form of open-ended questions. Teachers review students' written work and they are encouraged to provide feedback so that students deepen their understanding of the underlying mathematics. The *MathScape* curriculum provides students with many opportunities to express their ideas and justify their thinking. *Dr. Math* problems are designed to both probe and enrich student thinking through situations that are often related to common mathematical misconceptions (see Figure 17.2). Students assume the role of Dr. Math and respond to inquiries, as in the example.

Students must consider the context of the question and figure out what is the important and related mathematics. They explain in detail why something works and where the reasoning does not make sense. They take a position and support it mathematically. Tackling unique math problems through such open-ended questions helps students learn to write concise explanations about their mathematical thinking.

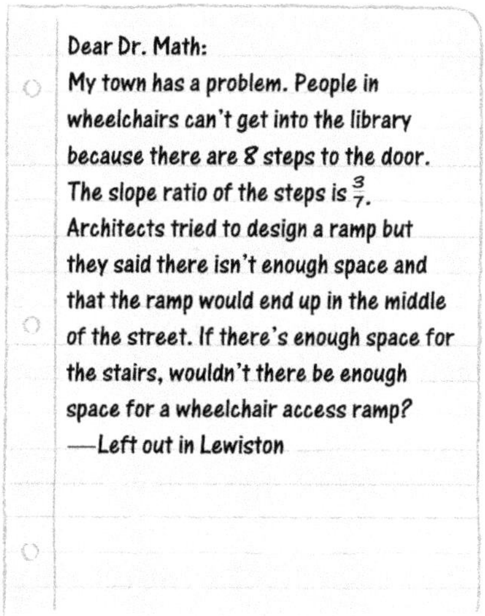
Dear Dr. Math:
My town has a problem. People in wheelchairs can't get into the library because there are 8 steps to the door. The slope ratio of the steps is $\frac{3}{7}$. Architects tried to design a ramp but they said there isn't enough space and that the ramp would end up in the middle of the street. If there's enough space for the stairs, wouldn't there be enough space for a wheelchair access ramp?
—Left out in Lewiston

Figure 17.2 An example of a *Dr. Math* problem (Brutlag, 2005).

DESIGN OF LESSONS AND UNITS

Each lesson of *MathScape* follows a consistent format. As teachers become familiar with the layout of the lessons, and see how content builds within and across the phases of a unit, they can evaluate a student's performance against specific learning goals. Standard elements of a *MathScape* lesson include: Opening, First Investigation, Processing/Reflection, Second Investigation, Closing and Reflection, and Practice and Extending.

Opening the Lesson

The teacher provides brief context and background to prepare students to complete the first investigation. The aim is to introduce the context in an engaging way and pose guiding questions to frame the first investigation.

First Investigation

The initial investigation is designed to begin with an interesting and accessible entry point for all students with potential for different outcomes. Students work in pairs or small groups to conduct an investigation in which they build and apply skills related to a particular context. This investigation prepares them for more advanced problem solving and analysis in the subsequent investigation. For example, a lesson on statistical measures begins with an initial investigation asking students to determine mean, median, mode, and range given a date table of basketball scores for three players.

Processing/Reflection

Typically the first investigation is followed by a class discussion facilitated by the teacher. Using guiding questions provided in the teacher's guide, the teacher promotes a discussion about the important mathematics from the first investigation. The discussion helps students in solidifying and synthesizing information and provides the teacher with an important opportunity to assess understanding at the midpoint of the lesson and prior to the next, more sophisticated investigation.

Second Investigation

The first investigation, with its accessible beginning, builds confidence in students and prepares them for the second investigation. In the second investigation students are called upon to perform more challenging mathematics—either by taking concepts learned in the first investigation and extending them, analyzing them, or applying them to different situations. This part of the lesson often asks students to try different approaches, analyze results, and justify their reasoning. For example, in the second investigation of the aforementioned lesson, students apply their understanding of statistical measures to decide and defend their choice of which basketball player is "best."

Closing the Lesson

After the two investigations, students have an opportunity to consolidate and demonstrate their understanding and reflect upon their learning with a carefully designed activity that occurs through a discussion, written task, or both. This part of the lesson also serves the purpose of assessing students' mathematical understandings.

Practice and Extending

Each lesson has a homework assignment that provides practice and reinforcement of concepts, and also provides an opportunity for students to try to extend their thinking or apply ideas in new contexts. For this reason, students will see both familiar problems and unfamiliar problems in the homework. The homework assignments contain problems in three categories:

- Applying the Skills: gives students problems that enable them to practice skills that are developed in the lesson.
- Extending Concepts: requires students to apply ideas to new situations.
- Making Connections: provides problems that relate to another subject or context, offers extensions, or asks students to write about their ideas and solutions.

Overall Lesson Flow

MathScape lessons are designed to support students in moving from informal to a more formal understanding of mathematics. To support this development, the curriculum first builds conceptually, investigating the essential mathematical ideas before introducing formal vocabulary and procedures. Consider, for example, the development of concepts in geometric measurement: Students first use a model to learn what area is, and then in the next unit, they learn and apply the formula.

Knowing the structure of the lessons and units will help teachers support students in their learning without pre-teaching, unnecessarily supplementing, or instructing in other ways that might compromise the goals for learning.

ASSESSMENT

MathScape provides support for a range of assessment techniques. These include performance assessments provided at the end of the phase, final projects, writing tasks, homework, portfolios, self-assessment, and group assessments. In addition, each unit includes three or four skill quizzes to help teachers assess student understanding.

Assessment in the *MathScape* curriculum is divided into two major categories: formative and evaluative. Formative assessment provides teachers information about how well students are grasping specific concepts and mathematical "big ideas," in order to inform instructional decisions. Evaluative assessment allows students to demonstrate what they know and can do, for teachers to assess their performance.

In the formative category, pre-assessments, day-to-day assessments, and student self-assessments are examples of how the materials structure assessment for teachers. Each unit begins with a pre-assessment activity designed so that teachers can check for important prerequisites and determine what students might already know about the content in the unit. Once the unit is underway, teachers can use day-to-day assessment tools such as the "what to look for" guidelines from the teacher materials to gauge student understanding and determine what next instructional steps to take. Student self assessments highlight for students the important learning goals and allow them to monitor their progress relative to these goals.

In the evaluative category, each unit includes a range of assessments. Each unit includes three or four performance assessment tasks that allow students to demonstrate what they have learned. For example, students design simulations of real-world activities that demonstrate their knowledge of probability and statistics. In addition to performance assessments, students are expected to show their understanding through skill quizzes and tests. Another assessment option, the portfolio, is a collection of artifacts that offer a picture of a student's progress over time. This range of evaluative assessments allows students to demonstrate mastery in a variety of ways and allows teachers to similarly assess using a variety of tools.

Rubrics

Rubrics are an important component of assessment in *MathScape* that can be used to clarify learning goals and expectations and evaluate student performance. For students, reviewing the scoring rubrics before responding to a written assignment clarifies the expectations of the assignment. This reduces the amount of guesswork in determining what information is important to include in the written response. For teachers, rubrics provide the criteria to prioritize what is most important for students to know or be able to do (see Figure 17.3). Teachers use the rubrics as measuring sticks to assess a student's mastery of key concepts and skills—does the student fully understand, partially understand, or have no understanding of the general principles or important mathematics in the unit?

CONCLUSION

Professional development that centers on the "big mathematical ideas" and the central features and philosophy of *MathScape* will build a solid understanding of the curriculum, how and why concepts are connected, and ways that students can learn these concepts well. It can build teachers' confidence in their ability to engage their students, help students apply their knowledge in varied problem-solving situations, and make connections among mathematical topics. During the professional development forums, teachers learn about ways that *MathScape* makes mathematical concepts explicit through interesting contexts, and how it promotes mathematical discourse as a means of expressing knowledge and creative ideas. Using context and communication effectively, teachers enhance the learning experience for their students, while simultaneously assessing their students' progress in grasping and applying mathematical concepts.

ASSESSMENT RUBRIC

Does student work...

▪ provide a sophisticated explanation of how the group analyzed their data and the graphs to figure out which bag they had? ▪ calculate the theoretical probability and experimental probability accurately and go beyond the assignment in some way? ▪ rank the five bags from the best to the worst theoretical probability of picking a red cube and picking a green cube correctly? ▪ include a sophisticated explanation of how the bags were ranked? ▪ use examples from the data and graphs to make a hypothesis about which bag the samples were taken from?	▪ provide clear descriptions of strategies used to determine which bag of cubes the group had? ▪ calculate the theoretical probability and the experimental probability accurately? ▪ rank the five bags from the best to the worst theoretical probability of picking a red cube and a green cube correctly? ▪ provide explanations of how they ranked the bags? ▪ use examples from the data or graphs to make a hypothesis about which bag the samples were taken from?	▪ provide descriptions of the strategies used to determine which bag of cubes the group had with some confusion? ▪ calculate the theoretical probability and the experimental probability with some errors? ▪ rank the five bags from the best to the worst theoretical probability of picking a red cube and a green cube with some errors? ▪ provide explanations of how they ranked the bags with some errors? ▪ include a hypothesis about which bag the samples were taken from?	▪ show that they were unable to describe the strategies used to determine which bag of cubes the group had? ▪ calculate the theoretical probability and the experimental probability with significant errors or missing work? ▪ rank the five bags from the best to the worst theoretical probability of picking a green cube with major errors or missing work? ▪ not include an explanation of how they ranked the bags? ▪ not include a hypothesis about which bag the samples were taken from?
4 3 2 1 Goes beyond expectations	4 3 2 1 Meets all expectations	4 3 2 1 Meets some expectations	4 3 2 1 Falls below expectations

WHAT DOES THE DATA SAY? • ASSESSMENT A19

Figure 17.3 Scoring rubric for *MathScape* assessment (Brodesky, Brown, & Tobin, 2005).

CHAPTER 18

KEY FEATURES OF *MIDDLE GRADES* MATH*THEMATICS* FOR PROFESSIONAL DEVELOPMENT

Rick Billstein and Jim Williamson
The University of Montana, Missoula, Montana

Middle Grades MATH*Thematics* (Billstein & Williamson, 1999a) is a comprehensive mathematics curriculum for grades six through eight that emphasizes new mathematical content, new ways to help students learn the content, and new ways to assess what students know and are able to do. The vision of the curriculum cannot be realized, however, unless teachers understand and incorporate into their practices changes in what they teach, how they interact with their students, how they assess progress, and how they communicate with parents and communities. Teachers and administrators need to understand that once a curriculum has been adopted, the process of change has only just begun. The goal of this chapter is to briefly describe the unique components of the MATH*Thematics* curriculum that should be emphasized in professional development designed for teachers, administrators, and parents.

A Decade of Middle-School Mathematics Curriculum Implementation, pages 183–189

ESSENTIAL COMPONENTS OF PROFESSIONAL DEVELOPMENT

Middle Grades MATH*Thematics* was initially funded by the National Science Foundation (NSF) under the name *Six Through Eight Mathematics* (STEM). The goals of the program are to help all students develop their abilities to reason logically, apply mathematical skills to real-life situations, communicate mathematically, and feel confident in using quantitative and spatial information to make decisions. The following characteristics of MATH*Thematics* should be emphasized within a comprehensive professional development program.

Content

To ensure that the curriculum is broad and balanced at each grade level, the learner outcomes are organized in strands: *Number, Measurement, Geometry, Statistics, Probability, Algebra,* and *Discrete Mathematics.* The content within each strand emphasizes problem solving, critical thinking, and reasoning over rote procedural drill. The emphasis in the curriculum is less on the review of elementary topics such as whole number computation, and more on data analysis, proportional reasoning, algebra, geometry, and discrete mathematics. This design has implications for professional development related to *new content, computation,* and *algebra.*

New Content. Since MATH*Thematics* contains mathematical content that is not typically taught in grades 6–8, many teachers are faced with content they have never encountered before. Middle school teachers are often elementary certified and have limited mathematics training. Most teachers with this certification are familiar with number-related topics, but algebra, geometry, statistics, probability, and discrete mathematics offer new challenges and teachers often lack confidence teaching them. Professional development for these teachers should articulate the key ideas being developed and the connections among various concepts.

Computation. Middle grade teachers and parents are most often concerned about the coverage of number topics, especially computational procedures. Professional development must address the role of computation and how to best teach it. Teachers need to see how they can embed the skill practice in problem-solving activities without having to pass out supplementary drill and practice sheets.

Algebra. MATH*Thematics* gives extensive attention to algebra and functions. Teachers need to see how algebra is integrated throughout the entire three years of the program and how it prepares students for later mathematics courses. Teachers also need to develop a deeper understanding of algebraic reasoning which includes work with symbols, patterns, rates of change, and exploring linear and nonlinear relationships.

INSTRUCTIONAL APPROACH

Middle school is a time when students should be actively involved in doing mathematics that is new and meaningful, not just a period of review. To accomplish this, the instructional approach employed in MATH*Thematics* actively engages students in:

- investigating, discovering, and applying mathematics,
- using concrete materials to explore mathematical properties and relationships,
- working cooperatively,
- communicating ideas orally and in writing,
- using technological tools when appropriate, and
- integrating mathematical strands to solve real-life problems.

In MATH*Thematics*, the units are developed using a spiral approach that does not expect immediate mastery of a concept. Teachers need to be informed about the strengths of a spiral approach to concept development and what mathematics is appropriate at each grade level. They also need to know how and when mathematical concepts are developed in a program, and how the content fits together to prepare students for later work.

To achieve the curriculum goals, the mathematics content is presented mainly through inquiry-based thematic modules that motivate learning by developing mathematical ideas in relevant and meaningful contexts. There are eight modules per grade level and each module is designed to last four to five weeks. The mathematics content of each module includes cross-curricular connections to other content areas such as science, language arts, and social studies. Work with computational skills is included with ample opportunity to practice skills that are developed in the program. By its very nature, this approach requires more reading and writing than is typically found in mathematics textbooks. The professional development program must provide teachers with materials and strategies to use with students to help them with content area reading and writing.

Traditional mathematics curricula focus on the answer, not on the process for finding it. In MATH*Thematics*, the focus is on the process as well as the answer. The process emphasizes constructing mathematics, not reproducing it; interacting with other students, not working in isolation; and integrating mathematics, not partitioning it into small segments. To change to a process-oriented curriculum, the beliefs of teachers and school communities about what mathematics is essential for students to know and how they are to learn that mathematics need to change. Without background and support, teachers are not likely to see the need for careful concept building that allows students to develop strategies and thinking processes.

MATH*Thematics* actively involves students in doing and learning mathematics through the use of hands-on activities and other varied instructional techniques. The curriculum also incorporates different support materials such as graphing calculators, computer software, videos, and concrete models. Teachers need help in understanding the role activity-based lessons, manipulatives, and support materials play in lessons and in concept development. The role of cooperative learning in the program must be discussed because about 30% to 40% of the activities in the program involve cooperative learning.

Because this curriculum is much different than traditional mathematics programs, pacing is a major issue. Teachers must be convinced to trust the program and not bring out extra drill sheets and try to teach to mastery the first time a topic is introduced. In some cases the mathematics seems hidden to teachers due to the contextual setting. It is very easy to get off the suggested timeline, especially the first time teaching a thematic approach. It is important that pacing be kept in mind and it is important that this is stressed in terms of professional development. Every strand needs to be traced throughout each grade level so that teachers can see the complete development. The strand development from grade level to grade level should also be studied.

USE OF TECHNOLOGY

Calculators, computers, and web-based activities are tools that can help students make sense of mathematical concepts by allowing them to display results, to find patterns, to research ideas, and to ask and answer "What if. . . ?" questions. Technology activities that can be used to develop and reinforce mathematical concepts, solve problems, and bridge the gap between the classroom and the real world are included throughout the curriculum. The materials, however, are designed so that they do not depend on the use of technology. Teachers need to be aware of how to use technology and be trained to use all of the components including web-based activities and the resources available on CD. Professional development with technology must include a discussion of not only what technology is available but also what technology is appropriate to teach different concepts.

ASSESSMENT

Assessment in MATH*Thematics* is an integral part of the instructional process, rather than an add-on to it. Not only is assessment information drawn from instructional tasks, but the assessment tools themselves are designed to help students master concepts and develop skills. The primary goal of the assessment package is to improve student learning. It serves four major purposes:

- to monitor student progress in problem solving, reasoning, and communication,
- to assess student proficiency in content areas,
- to help teachers make instructional decisions, and
- to document student progress for students, parents, and teachers.

Because assessment in MATH*Thematics* is an integral part of instruction, many assessment items are embedded in the instructional materials. *Checkpoints, Discussion* questions, and *Try This as a Class* exercises are included to provide guidance to teachers on how and when to interact with students. Professional development in the area of assessment should include helping teachers with issues such as: How can you ensure students are given a balanced task? How do you score these tasks? What is the difference between holistic and task-specific scoring? How do you achieve the maximum in objective or subjective scoring? How does what you assess help direct your teaching? When should you assess skills and when should you assess student performance? How do you report results to parents?

Checkpoints are questions or problems that are used to check understanding of a concept or skill before students continue with an exploration. *Discussion* questions provide places for students to check their understanding of a concept by sharing or generating ideas within their group or as a class. *Try This As A Class* questions appear at points where direct instruction is needed to summarize key ideas or to bring closure to a line of inquiry. They are similar to *Discussion* questions, except that the teacher directs the discussion or activity and guides the learning.

Extended Explorations (E^2s), are typically open-ended or open-response problems that students complete independently—or occasionally in small groups—outside of class. E^2s apply a variety of mathematical concepts and may be solved in different ways. Each module contains one E^2. It usually takes one to two weeks for students to complete an E^2, and their solutions are assessed using a multi-dimensional generalized assessment rubric.

Assessment Rubrics are a key to raising student performance by actively involving students in assessing their own work. This is achieved through use of a generalized multi-dimensional *Student Self-Assessment Rubric.* The rubric scales were designed to help students answer the question "How can I improve my problem-solving skills, reasoning, and communication skills?" Five scales are included: *Problem Solving, Mathematical Language, Representations, Connections,* and *Presentation.* Combined, they provide a generalized rubric that defines the various dimensions of mathematical investigation. As students become familiar with the scales, they understand what they need to do to improve their problem-solving, reasoning, and communication skills.

Teachers assess students' work using a *Teacher Assessment Rubric.* The teacher scales are similar to the student scales but are designed from a teacher's point of view. The assessment scales allow the teacher to recognize a student's strengths and help the student focus on areas where growth is needed. The combination of student and teacher assessment provides important feedback to help students improve. Portfolios are used to provide comprehensive documentation of students' progress in, attitude toward, and understanding of mathematics over a period of time.

Modeling assessment with the generalized multi-dimensional scoring rubrics should be a major focus of professional development. Teachers must be involved in doing self-assessment, peer assessment, and teacher assessment before they actually try them with students. Teachers need to be familiar with the assessment scales and feel comfortable enough that they can use the scales to grade student work without taking a long time to do it.

DELIVERING PROFESSIONAL DEVELOPMENT

Teachers need to become familiar with the philosophy, organization, and components of the MATH*Thematics* curriculum during professional development. They need to engage in hands-on activities to closely examine the key components and mathematics content of the program. Opportunities to explore the mathematical content, to engage in assessment activities, and to model parent involvement, lesson planning, module planning, and project planning must be included. To model the program's philosophy of teaching, activities should be presented as they would be taught to students. It is important that teachers work through as many activities as possible before teaching them. In this way the teacher can discover what approaches a student might use and will be able to help students when they have questions.

The trainers in the professional development workshops are often lead teachers who have undergone extensive training with the developers and/or McDougal Littell consultants. If these trainers come from within a district, they can be available to assist in district in-service during the year. An important part of professional development is that teachers have opportunities to talk with other teachers about content and teaching issues. It is also important that teachers have follow-up training during the year as questions arise.

The length and format of staff development can vary depending on the needs of the school. Professional development can be customized to meet the needs of the school and its staff. A *Professional Development Handbook* (Denny & Williamson, 1999) that comes with the program is available for use with each grade level. A sample schedule for an introductory 2-day teacher in-service and 1-day trainer workshop is given in Table 18.1.

TABLE 18.1 Sample Schedule for MATH*Thematics* Professional Development

2 Day Teacher In-Service and 1 Day Trainer Workshop

Schedule		
Initial In-Service	*Teachers and Trainers* 2 full days	*Trainers* 1 full day
Follow-up In-Service	*Teachers and Trainers* 1 after school session	*Trainers* 1 full day

Day 1 of Teacher In-Service (6 hours)

Introduction
Changes in Mathematics

*MATH*Thematics
Philosophy
Organization and Structure
Strands
Materials List
Questions and Answers

Close-up
MATH*Thematics* Section
Setting the Stages
Explorations
Key Concepts
Practice and Application Exercises
Projects
Teacher Resources
Technology
Questions and Answers

Instruction
Problem Solving
Cooperative learning
Extended Explorations
Questions and Answers

Day 2 of Teacher In-Service (6 hours)

Assessment
Purposes of Assessment
Assessment Tools
Assessment scales
Questions and Answers

Communicating with Parents and Community
Newsletters
Math for Parents
*MATH*Thematics for Administrators
Questions and Answers

Lesson Planning
How to Plan a Daily Lesson
Key Components
Teaching Tools

Unit Planning
Planning Full Modules

Special Topics
Algebra in *MATH*Thematics
Computation in *MATH*Thematics
Content Area Reading and Writing Strategies
Questions and Answers

Full Day Trainer Workshop (6 hours)

Extended Planning
Assessment
Lesson Planning
Unit Planning

Instruction
Adjusting for Special Needs
Scope and Sequence
Questions and Answers

Follow-up In Service (mid-year)

Full Day Trainer Workshop
Cooperative Learning Strategies
Review/Extend
Extended Explorations
Assessment
Unit Planning
Program Evaluation
Questions and Answers

After School Teacher In-Service
Questions and Answers
Cooperative Learning
Unit Planning

CHAPTER 19

TEACHERS' NEEDS AT DIFFERENT PHASES OF CURRICULUM IMPLEMENTATION

Fran Arbaugh[1]
University of Missouri, Columbia, Missouri

As the authors in this section have made clear, professional development for teachers implementing a NSF-funded curriculum is critical. For most teachers, this type of curriculum is vastly different from what they have been using in their middle grades mathematics classrooms; students actively participate in learning mathematical concepts and teachers work as a facilitator and organizer. Experience, as well as research, tells us that for successful implementation, much attention must be paid to teachers' professional growth as they "relearn" to teach middle grades mathematics.

The Concerns-Based Adoption Model (CBAM) (Hord, Rutherford, Huling-Austin, & Hall, 1998) provides a useful framework for us to consider the needs of teachers as they adopt and implement a NSF-funded mathematics curriculum. These researchers argue that teachers undergoing change are concerned about different issues at different points in the change process:

A Decade of Middle-School Mathematics Curriculum Implementation, pages 191–199

- Self-concerns: What is this new change and how will it affect me?
- Task-oriented concerns: How do I implement this change? What do I need to do to make this change happen with my students?
- Impact-oriented concerns: How are my students learning? Are they learning more and are they learning better? How do I work with others who are also implementing these new ideas?

Professional development for teachers initially implementing a NSF-funded middle school curriculum needs to be different from the professional development for teachers who have been implementing one of these curricula for a number of years. Goldsmith, Mark, and Kantrov (1998) call the type of professional development needed at the time of initial implementation "Meeting 'Just in Time' Implementation Needs." They contend that teachers have very specific needs that must be addressed as they begin to implement a curriculum. Those needs change as teachers become more adept at handling classroom management issues, pacing concerns, and grading issues. We encourage leaders in charge of professional development to consider the different needs of teachers at different stages of concern, and design appropriate professional development opportunities for the teachers.

The purpose of this chapter is to support leaders, charged with the professional development of teachers using these type of curricula, to design learning experiences for teachers to address concerns during three stages of implementation. These stages of curriculum implementation occur when teachers are:

1. learning about and adopting;
2. initially implementing; and
3. a few years into implementation.

TEACHER CONCERNS AT THE ADOPTION STAGE

Before teachers begin to examine a NSF-funded middle grades curriculum, they should engage in discussions about mathematics teaching and learning as described in the *Principles and Standards for School Mathematics* (National Council of Teachers of Mathematics, 2000). Teachers need to have a reason to be interested in adopting a new kind of curriculum. If they are satisfied with their current curriculum, they may not see a reason to change.

As teachers begin to engage in examining these curricula, their concerns typically are of the *self concern* type. For example, at the Show-Me Center, we have typically heard questions about:

- the supplementary resources that are available with the curricula;

- how much time it takes to plan for a lesson using a particular curriculum;
- how much time it takes to grade homework/quizzes/tests;
- the availability of manipulatives (and the willingness of their districts to buy them);
- why these curriculum materials "look" so different from the textbooks they have been using in their mathematics classes; and
- the role of the teacher.

Leaders of professional development must be responsive to the "self concern" needs at this stage of use. Teachers are justifiably concerned about trying such a different approach to teaching mathematics.

The most successful adoptions occur when teachers are given ample time to get to know these textbooks and supplementary materials. Several venues exist for teachers to build knowledge with the support of others more familiar with the curricula. One good source of information about the four curriculum projects is the Show-Me Website (http://showmecenter.missouri.edu). Note that the website contains examples of activities for each curriculum accompanied by teacher pages and examples of student work. Further, some of the activities are linked to the MMM Videos (Bolster Education, 2001), allowing visitors to see into classrooms where these activities have been implemented. A second venue is through the publishers of the curricula. In most cases, publishers encourage teachers and districts to pilot their materials.

TEACHER NEEDS AT THE STAGE OF INITIAL IMPLEMENTATION

Once a district/school adopts one of the NSF-funded curricula, teachers transition into an "initial implementation" stage, which can last up to three years past adoption. *Self concerns* are still prevalent at this stage of adoption, but teachers also begin to ask questions that reflect *task concerns* (e.g., How can I implement these materials in my classroom? How do I manage manipulatives?)

At the beginning of this stage, and prior to the first year of implementation, opportunities must be provided for teachers to "get into" their adopted curriculum. Teachers in this stage typically express concerns about a number of issues, and the work done just prior to first-time implementation can be very beneficial for teachers and their students. Teacher concerns at this time typically focus on classroom-based issues at several phases of instruction: planning, classroom implementation, and assessment.

The Planning Phase

During the initial implementation stage, teachers often spend many hours getting to know their curriculum and planning for instruction. The following two areas are often of special concern to teachers during planning.

- *The Path of Mathematical Strands.* During this stage, teachers often do not have knowledge of the direction a particular strand takes through the grades 6–8 curriculum. Teachers in lower grades often do not know the future direction of the work they are doing with students, while teachers of the higher grades often do not know to what extent students have learned the necessary prerequisite knowledge. Once teachers see this progression of mathematical concepts within the mathematical strands, they better understand the content, as well as where the content occurs. Supporting teachers at this stage to know the "big picture" of the curriculum helps with these concerns.
- *Lesson Planning.* Lesson planning is another area that often takes a large part of teachers' time when they begin implementing a standards-based curriculum. During the initial implementation stage, it is helpful for teachers to plan collaboratively by working through lessons and modules or units, emphasizing the mathematics of the lessons. This initial planning is well worth the time investment, especially as it helps teachers see how much time might be involved for students to work through the materials, either in class or as a homework assignment.

The Classroom Implementation Phase

Teacher concerns shift a bit when they start teaching the curriculum. The following areas are often of concern to teachers during the classroom implementation phase:

- *Mastery vs. Spiraling of Skills.* One major area of teacher concern involves mastery vs. spiraling of skills. Many teachers in the initial implementation stage feel the need to pre-teach skills, or spend extra time drilling until all students have mastery. However, these curricula are designed so that the spiraling of skills over the year, as well as through the entire middle school program, leads to mastery. The practice of teaching to immediate mastery has a negative effect on pacing. Many teachers find that they are moving so slowly that they are not able to progress through the curriculum in order to see the mastery developing through repeated contact with a skill or mathematical strand.

- *Materials Management.* At this stage, teachers can also benefit from advice about materials and classroom management. For example, what is a good way to store and distribute manipulatives? What is a good way to begin class each day? What expectations and procedures are there for absentees and make-up work? Teachers will want help and answers for these common questions about the day-to-day workings of their classroom.
- *Instructional Strategies.* Teachers also have questions about the teaching of the lessons and explorations or investigations. These questions may concern inquiry-based learning, and what types of questions prompt student thinking. Teachers have questions about how lessons look in terms of independent investigation versus teacher-directed lessons. For example: How much should the teacher lead, if at all? How do we develop a student's confidence in his own thinking, and in the thinking of his peers? Much of this involves the bigger question of how students learn, as well as how teachers teach.

Assessment Phase

Many questions arise concerning assessment, as it is often very different than in traditional textbooks. The following are often of concern to teachers during the assessment phase:

- *Assessing Daily Work.* Many teachers need support as they make decisions on assessing daily work. These curricula require that students actively work on mathematics on a daily basis, and teachers often become overwhelmed with the amount of paper generated. Teachers need to have a realistic plan for processing class-work and homework.
- *Using New Types of Assessments.* These curricula offer many alternative methods of assessment. If teachers use the assessments provided, questions often arise about how to grade assessment problems, how to assign letter grades to alternative assessments, and how to manage their time when grading assessments.
- *What are Students Learning?* Teachers also need to think deeply about what the assessments tell them about what their students are learning. Teachers are presented with a plethora of data about what students know and do not know. Often teachers are overwhelmed with all of the data. They begin to think about using assessment information for more than having a grade to put in a grade book.

> Ultimately, teachers need support to be able to answer this question: How well does the student know the mathematics?

Teachers in the stage of early implementation need to be supported with time to meet with colleagues to work through upcoming units, share assessments, discuss student work, debrief on instruction, and generally provide each other with collegial support.

TEACHER NEEDS DURING CONTINUING IMPLEMENTATION

After about three years of initial implementation, teachers transition into the next phase: continuing implementation. It is at this stage that opportunities for continued growth tend to diminish—often just when their concerns become more impact-oriented. Teachers know the curriculum better than when they first began using it; many classroom and assessment managerial issues have been solved; students and parents are getting used to the new curriculum.

Although it may appear that there is no need for continued professional development for teachers in this stage, this is the optimal time for teachers to engage in opportunities that encourage them to reflect deeply on their instructional practices and students' learning. Typically, issues of concern for teachers in this phase are more reflective in nature and, in some cases, teachers may have difficulty identifying their own needs—they just know that they "need something more." Professional development takes on a different look at this stage; instead of putting more "tools in the toolbox," we need to provide opportunities for teachers to work collaboratively to refine the tools they already have and to engage in a deep study of their classroom practices.

This kind of deep study can be supported by offering teachers a framework within which to couch their learning. For example, Hiebert, Carpenter, Fennema, Fuson, Wearne, Murray, Human, and Olivier (1997) describe five dimensions of classrooms that promote mathematical learning:

1. A classroom culture that honors inquiry;
2. A lesson/task that is mathematically worthwhile and problematic;
3. Tools that are totally integrated to develop a depth of understanding;
4. Teachers who support students' construction of knowledge and understanding; and
5. Classrooms where all students have access to and opportunity to learn meaningful mathematics.

Some issues, such as teaching reading in mathematics, helping students who struggle with writing explanations, managing cooperative groups, and developing the increased ability to ask probing questions, are often a focus of initial implementation professional development. However, teachers at the continuing implementation stage begin to want more in-depth information and specifics as well as the opportunity to work with other teachers around these issues.

A larger issue that confronts teachers as they move further into implementation is that of mathematical content knowledge. In some cases, the teachers recognize a need for deeper understanding or for making connections not previously made in their mathematics background. In other cases, teachers' understanding of the mathematics is on a surface level, and they do not realize the missed opportunities for enhancing student understanding, therefore limiting the student learning potential embedded in these curricula.

After teachers are more comfortable with the logistics of implementation, they are more comfortable with working collaboratively in and out of the classroom. Collaboration between mathematics teachers and special educators begins to emerge during later stages of implementation. In initial unit seminars, teachers tend to focus only on the grade at which they are teaching. This limits their ability to communicate with other teachers in their school. As the implementation progresses, teachers become concerned about developing a better understanding of what happens before and after their grade thus moving away from "What do I do?" toward "how are *we* helping children learn mathematics?"

It is important to note that even if a district has moved to the assessment phase in its implementation, new teachers (and teachers new to the district) will probably be in the planning and implementation phases. Professional developers need to always be aware and plan for educating new teachers about the curriculum and then supporting their needs during implementation.

Experienced users of a NSF-funded curriculum often become teacher leaders in a district, across a state, or on a national level. These teacher leaders can be invaluable in educating new teachers (who are in the planning phase) about the curriculum and in supporting their peers (who may be in the implementation phase or the assessment phase). Many times these teacher leaders seek ways to increase their own knowledge about issues in mathematics education (e.g., seeking advanced degrees; attending national teacher meetings; reading professional literature). The challenge for a district professional development coordinator is twofold: (a) how to support the new educational needs of these teacher leaders; and (b) how to justify taking successful teachers out of the classroom and shifting their responsibilities to a more global role such as mathematics coaches or mentors.

CONCLUSION

This chapter has detailed how the adoption and implementation of a NSF-funded curriculum provides an opportunity for significant teacher growth. It provides a structure for those individuals in a school or district who organize professional development around curriculum adoption and implementation. The two main points that district-level professional developers should recognize are that teachers have different needs at different times in the implementation and that professional development must be ongoing for the life of the implementation.

Missing from this chapter thus far is discussion of the impact of teacher turnover with regard to teachers' needs during NSF-funded curriculum implementation. Few districts understand or anticipate the disruption caused by teacher turnover once these types of curriculum materials are in place. Teachers leave and are replaced, schools expand and hire new teachers, teachers change content areas or grade levels. By the third year of implementing a new curriculum, the make-up of faculty can be quite different than it was during the initial years of implementation.

We live in a mobile society and are accustomed to a changing staff of teachers. Teachers new to a district will require the same types of professional development discussed in this chapter. This is critical because teachers new to the school are unlikely to have taught from one of these programs or have learned in one of these programs. Typically, a new teacher is handed a textbook and a curriculum guide and begins teaching. This process has proven to be problematic when the teacher is new to a NSF-funded curriculum.

Bay (1999) found that in a district with little turnover, after three years of implementation, a commonly accepted rationale existed for implementing one of the NSF-funded curricula; further, the curriculum was perceived positively by teachers, administrators, and parents. Conversely, in another district that replaced about a third of its teachers over a three-year period, there were mixed opinions about why the materials were in use, including those of teachers who had no knowledge of the decision making that had gone into the adoption decision. In addition, the teachers who did not understand why the curriculum materials had been selected were unaware that the curriculum used a different instructional approach and therefore tried to teach the curriculum in a traditional manner of teacher explanation-student practice-homework. As experience has shown, if a school has a mobile teacher population and does not provide targeted curriculum support to incoming teachers, the result can be a poor quality implementation or abandonment of the program in the next adoption cycle.

Sustaining mathematics curricular reform involves on-going support for those teachers at many stages of implementation. It also involves special professional development targeted for those teachers who arrive after the initial implementation. Providing support for teachers should be at the forefront in a district's plan for successfully adopting and implementing a NSF-funded middle school mathematics curriculum.

NOTE

1. A previous version of this chapter was written by Fran Arbaugh, Murrell Hoover, Linda Tetley, and Dustin Jones.

CHAPTER 20

PROFESSIONAL DEVELOPMENT FOR ADMINISTRATORS

What They Need to Know to Support Curriculum Adoption and Implementation

Margaret R. Meyer
University of Wisconsin–Madison, Madison, Wisconsin

and Fran Arbaugh
University of Missouri, Columbia, Missouri

> *...place the issue of administrative support first. Indeed, having an administrator take the time from day one to participate in and learn about these curricula had an enormous impact on teachers' continued efforts to adopt and implement a standards-based curriculum.*
>
> —Bay, Reys, & Reys, 1999, p. 504

There is little debate about the central role of administrators in supporting standards-based mathematics curriculum adoption and implementation. Several chapters in this volume clearly demonstrate the importance of strong administrative support and leadership. However, it is also recognized that the people who most need to be involved in supporting curricu-

A Decade of Middle-School Mathematics Curriculum Implementation, pages 201–210

lum adoption and implementation are the ones who often have the least amount of time and knowledge to devote to the task.

In this chapter, we describe a workshop aimed at providing minimal key information about curriculum adoption and implementation to school administrators. It is expected that the reader who is interested in using this workshop will see it as a beginning and tailor it to their local needs.

This workshop was designed to include a two-hour session in the morning and another two-hour session in the afternoon, separated by an hour for lunch and conversation. The goals of the workshop are:

- to expose administrators to standards-based middle grades math curricula
- to help administrators recognize the work to be done in choosing a curriculum and in supporting its implementation
- to help administrators know their role in this work
- to provide additional resources administrators can use beyond the workshop.

The focus of the morning session is on understanding the curricula better. The afternoon session focuses on the role of administrators in curriculum adoption and implementation.

WHAT ARE THESE TEXTBOOKS, ANYWAY?

As with professional development sessions designed for teachers, we suggest that the day begin with introductions, followed by a preview of the day's activities. Once the format of the day is established, engage the administrators in Activity 1: Examining Curricula. This activity works best when the participants are grouped at tables to allow easy examination of materials from the four curricula as well as to support collaborative discussion about what they find.

Activity 1: Examining the Curricula

As an introduction to the Examining Curricula activity, we found that a few key questions get participants to begin thinking about the task:

1. Why are you looking for a new curriculum? Most administrators have very clear reasons why they are looking for a new middle grades mathematics curriculum, although sometimes they have not articulated these reasons. It is useful for the workshop participants to hear

that other people in their position share their reasons. Allow a few minutes for quiet reflection and then share out the reasons in a whole-group discussion. You can decide if it is helpful to record the reasons on a flip chart or transparency. Some of the reasons most often given are:
  - wanting to raise test scores
  - wanting to jump-start professional development of the teachers
  - wanting students to be better prepared for high school mathematics
  - needing a better alignment with the state curriculum standards
  - wanting all students to be successful with mathematics.
2. What is different about NSF-funded curricula? The answer to this question is less clear to most administrators. This is the first opportunity many of them have had to examine the NSF-funded materials themselves. Have available on each table sample materials from the four curricula: *Mathematics in Context* (National Center for Research in Mathematical Sciences Education & Freudenthal Institute 1997, 1998; Wisconsin Center for Education Research & Freudenthal Institute, 2006) *Connected Mathematics Program* (Lappan, Fey, Fitzgerald, Friel, & Phillips, 1998, 2002, 2004, 2006), MATH*Thematics* (Billstein & Williamson, 1999a), and *MathScape* (Education Development Center, Inc., 1998, 2005). Invite them to flip through the materials with the following discussion questions to guide them:
  - What do you see?
  - What challenges do you anticipate?
  - What are the potential benefits?

Administrators typically see many things when they do the "flip test," but not always the most important things. One of the first things they see is fewer problems and more words. They need help to see that the quality of the problems is different then those found in more traditional textbooks; this can be accomplished using an activity described in a later section (Activity 2: Doing a Mathematical Task). The increased need for children to read is easily seen as both a challenge and an opportunity. Most administrators will see that the materials are full of contexts that are designed to be engaging and motivating. A deeper look is required to see that the contexts serve as a way for students to mathematize their worlds and to anchor their thinking. Administrators also recognize that parents might have questions and concerns about the materials and they begin to think about how to prepare them for the changes.

After the administrators have had time to examine the materials, orchestrate a whole-group discussion around the two questions above.

Activity 2: Doing a Mathematical Task

After the administrators have a chance to talk about what they see from a surface examination of the materials, it is time to get them to take a deeper look by experiencing a problem or activity from one of the curricula in much the same way that the middle school students will. There are several criteria to keep in mind when choosing a problem or task for this activity:

1. The problem should require thought—that is, the solution should not be obvious. On the other hand, it should not be so hard as to make the administrator feel anxious or inept.
2. The problem should reflect significant mathematics; often including mathematics that is not typically done in middle school.
3. The problem should allow for multiple solution strategies, some at a more informal level and some at a more formal level. A problem that produces differences in solution paths is very good in that it requires people to justify their solution.
4. The problem should stand somewhat alone. Oftentimes in these materials a problem context builds over several pages and it is difficult to complete one question in isolation.
5. The activity or problem must fit the time available, leaving time for discussion and questions.
6. A problem that is hands-on is a good choice. However, you do not want to make the administrators feel like everything in the curriculum is of this nature.
7. Have student work to show whenever possible. This makes the learning by middle grades students believable.
8. Whenever possible, involve a strand of mathematics besides Number and Operations so that the administrators see that middle school mathematics is more than just number.
9. Finally, be sure that you are comfortable guiding the activity and modeling appropriate pedagogy.

It is difficult, if not impossible, to select a problem that meets all of the above criteria. However, a well chosen problem will provide a lot of fuel for a discussion of the following questions:

- *What is the mathematics?* A good problem will have mathematics from multiple strands and produce connections between big ideas. It will also suggest to administrators that some of their teachers might need support to learn the mathematics themselves before teaching it to their students.

- *What is the work of the teacher?* The answer to this question is a reflection of the way in which the problem is facilitated in the workshop. Facilitation should model active guiding with little teacher talk.
- *What is the work of the students?* As administrators reflect on the work of the students, they should realize that classrooms using these curricula are likely to be noisy, as students talk to each other while solving the problem and then when presenting their strategies and justifying their solutions. This may prompt the administrators to consider that they might have to re-evaluate how to observe or "walk through" a classroom of this sort.
- *Are the materials "teacher friendly" or "teacher serious"?* To answer these questions, the administrators must be able to see what is in the Teacher Guide. Hopefully they will see that the Guide provides the necessary support for teachers to engage students in the problem or activity. Addressing this question should also suggest to the administrators that teachers need to have ongoing professional development around the curriculum as well as planning time with other teachers.

Morning Session Summary

By way of summary, present the following characterization of NSF-supported curricula:

- Comprehensive programs, not supplementary or ancillary
- Designed to reach *all* students
- Focus on investigations that interest, challenge, and motivate students
- Identify and explore "significant" math
- Prepare students for math in high school
- Engage students as active learners
- Require different roles for teachers and students
- Illustrate different forms of assessment
- Support professional development opportunities for teachers

PREPARING THE WAY FOR MAKING A CURRICULUM CHOICE

Administrators attending a workshop of this type will typically be at many different stages in the adoption process. For some the decision will have already been made, while for others the process is just beginning and they can benefit from the experiences of others. Perhaps the adoption stories in section 2 of this book contain important insights. This topic could easily merit a full-day workshop of its own for administrators. However, in this workshop a few key points deserve attention.

End the morning session by presenting the following issues for administrators to consider about the adoption process in their school or district. As you present the issues, encourage participants to contribute their insights. Encourage discussion of these issues during the "working" lunch.

- *Decision making: Top down or bottom up?* Adoption decisions made by teachers potentially have very different outcomes than those made solely by administrators or school boards. Teachers usually feel like a curriculum has been forced upon them if they do not have a significant role in the decision process. However, it is difficult to achieve buy-in from all teachers even when they are involved in the process. The adoption process should begin early enough to allow input from all stakeholders. But even in this situation, administrators need to recognize that their leadership is a critical factor in the adoption process.
- *Review local frameworks and assessments:* Many districts spend considerable time trying to match their local frameworks and assessments to the available curricula. It is hard to argue against this attention, except in the case where the framework is in need of revision or is overly specific and restrictive. For example, a framework that stipulates that a certain mathematical topic be taught in grade 6 might preclude consideration of a very good curriculum that places that topic in grade 7. Likewise, the match between assessment and curriculum is important. If the local or state assessment includes problem solving and open-ended problems, children need a curriculum that provides opportunities to learn how to respond to these problems. On the other hand, an assessment that focuses only on recall knowledge and computational fluency will not be able to measure the extent of students' learning with one of the four standards-based curricula.
- *Setting criteria:* What are you going to look for when examining curriculum materials? There are many possible criteria: publisher commitment to providing professional development, content alignment with state and local frameworks, teacher support materials, affordability, durability, and so on. Establishing and distributing criteria before the review process helps districts avoid curriculum decisions based upon non-essential characteristics like the number of pretty pictures or the publisher promise of test-prep software. A good resource for criteria to consider is: Goldsmith, L.T., Mark, J., & Kantrov, I. (1998). *Choosing a Standards-Based Mathematics Curriculum.*

Newton, MA: K–12 Mathematics Curriculum Center, Educational Development Center, Inc.

- *Visiting classrooms in nearby districts:* Many districts like to be able to send a group of their teachers to nearby districts where a particular curriculum is being taught. In theory, this is a good idea because it allows teachers to talk to other teachers about what it is really like to teach with a particular curriculum. However, the administrator planning such a visit should be sure that the district the teachers are visiting has a strong implementation of the curriculum.
- *Making sense of a pilot:* Many districts decide to pilot one or more units from the different curricula. This is a good tactic to build teacher buy-in, but it has to be done carefully or the information generated by the pilot may not be helpful. For example, if one group of teachers pilots curriculum ABC and another group pilots curriculum XYZ, how will they come to consensus about which curriculum to choose? One way to design a pilot that produces sharable information is to have all teachers pilot the same unit at the same time. This model for piloting will allow for common planning and debriefing, and a common experience. The curriculum developers can help districts identify a representative unit that can be taught at multiple grade levels.
- *Professional development:* District administrators need to begin thinking about professional development well before the adoption process. A key question to ask of the publisher of a curriculum under consideration is the amount and type of professional development required and the role of the publisher in providing it. Administrators need to be realistic about the amount of professional development a publisher can provide as part of the "sales package." A large urban district that spends millions of dollars to buy materials for thousands of students will be in a much better position to negotiate for professional development than a district that spends a few thousand dollars for materials for a few hundred students. The teachers in these two districts have the same professional development needs, but it is just not economically feasible for a publisher to provide the same level of service to both districts. Encourage the administrators at the workshop to talk about how smaller districts can provide startup and ongoing professional development for its teachers.
- *Ask about achievement data:* Another common request from administrators when examining curricula is achievement data, the reasoning being that if the curriculum works in other districts, it will work in mine. This assumption fails to consider many other factors that contribute to successful implementations such as differences between the districts and the amount of professional development,

the commitment of the teachers, the support of the parents, the number of years using the curriculum.

OK, WE MADE A CHOICE. NOW WHAT?

The second half of the workshop leads administrators to explore what happens after a curriculum is adopted and they prepare for the first year of implementation.

Activity 3: Who is Responsible for What?

The intention of the activity that follows is to answer two questions:

1. What is the role of major stakeholders in curriculum implementation?
2. What does each group need to meet these responsibilities?

Divide the participants into equal-sized groups representing the following stakeholders: Teachers, Students, Parents, and Administrators. Provide each group with one or two sheets of flip-chart paper and markers. Give them 10–15 minutes to identify and record the responsibilities their group has for the success of the curriculum implementation. Some possible responsibilities for each group might include:

Teachers:
- Attend professional development sessions to learn about the curriculum.
- Plan units and individual lessons.
- Communicate expectations to the students and their parents.

Students:
- Come to class prepared to learn.
- Do the homework.
- Attend school regularly.

Parents:
- Support the efforts of the teacher.
- Monitor their child's homework.
- Attend parent information sessions.

(Participants will enjoy identifying the responsibilities of students and parents).

Administrators:

- Protect teachers from outside pressures.
- Provide common planning time for grade level teachers.
- Support ongoing professional development.

Circulate around the room and answer any questions the groups might have. Provide additional paper as needed. End the activity when the administrators seem to be running out of new ideas. Post the lists on different walls of the room leaving adequate room between postings of different groups. Now have the participants move in their group from list to list, adding additional responsibilities to the list using a different colored marker. Have everyone move in the same direction to minimize traffic snarls. Allow adequate time for reading each list and adding responsibilities: about 3–5 minutes per list. Then move on to the next list. At the end each group should revisit their own list to see what was added.

Wrap up the activity with a general whole-group reflection. What did they see? Any surprises? The two most important messages the administrators should go away with from this activity is that as administrators, they have a lot of responsibilities for the success of the implementation, but they are not responsible for everything. This message sets up the final activity of the workshop which is a period of final reflection and planning.

Activity 4: The Role of the Administrator

Begin this final activity by having participants write quietly for 5 minutes using the following prompt: *As an administrator involved in a curriculum implementation, I need to provide support for:*

- *Teachers*
- *Students*
- *Parents*
- *Myself and my administrative staff*
  in the following ways...

After the administrators have finished writing, ask for volunteers to share their reflections with the other administrators at their table. As they finish at their tables, ask if anyone wants to share with the whole group.

Workshop Closing

We have found that an inspiring way to close this administrator workshop is to distribute copies of the following quote by Alfie Kohn:

> Finally, whatever your position on the food chain of American education, one of your primary obligations is to be a buffer—to absorb as much pressure as possible from those above you without passing it on to those below. If you are a superintendent and must face school board members who want to see higher test scores, the most constructive thing you can do is to protect principals from these ill-conceived demands—to the best of your ability and without losing your job in the process. *If you are a building administrator, on the receiving end of test-related missives from the central office, your challenge is to shield teachers from this pressure—and, indeed, to help them pursue meaningful learning in their classrooms* (emphasis added). If you are a teacher unlucky enough to work for an administrator who hasn't read this paragraph, your job is to minimize the impact on students. Try to educate those above you whenever it seems possible to do so, but cushion those below you every day. Otherwise you become a part of the problem. (Kohn, 2001, p. 351)

ADDITIONAL RESOURCES

We end this chapter with a list of additional resources that may be useful to administrators throughout the process of adoption and implementation of NSF-funded middle grades mathematics curricula.

The following resource list should be distributed to the participants. If your workshop budget allows, obtain books for each participant or make copies of the key articles.

Bay, J. M., Reys, B. J., & Reys, R. E. (1999). The top 10 elements that must be in place to implement standards-based mathematics curricula. *Phi Delta Kappan, 80*(7), 503–506.

Goldsmith, L.T., Mark, J., & Kantrov, I. (1998). *Choosing a standards-based mathematics curriculum.* K–12 Mathematics Curriculum Center, Educational Development Center, Inc, Newton, MA.

Leinwand, S. (2000). *Sensible mathematics.* Portsmouth, NH: Heinemann.

Meyer, M. R., Delagardelle, M. L., & Middleton, J. A. (1996). Addressing parents' concerns over curriculum reform. *Educational Leadership, 53*(7), 54–57.

Mirra, A. (2003). *Administrator's guide: How to support and improve mathematics education in your school.* Reston, VA : National Council of Teachers of Mathematics.

Reys, B., Chavez, O., & Reys, R. (2003). Middle school mathematics curriculum: A guide for principals. *Principal Leadership, 3*(7), 61–66.

CHAPTER 21

A STORY OF TEACHER GROWTH

Increasing Expectations

Jane Peterson Porath
East Junior High, Traverse City, Michigan

During the spring of 1996 I was a student teacher at Traverse City East Junior High. At that time, Traverse City Area Public Schools was involved in a pilot with the *Connected Mathematics Project* (*CMP*) (Lappan, Fey, Fitzgerald, Friel, & Phillips, 1998, 2002, 2004, 2006). This was my first experience with a curriculum like *CMP* and I was amazed at how differently mathematics could be taught and learned. I remember wondering if this "new" approach to mathematics was a complete revolution or more of a baby-step toward what was yet to come.

During that semester of student teaching, I worked with a group of 8th-grade students. There was one experience with the unit *Moving Straight Ahead* (Lappan, Fey, Fitzgerald, Friel, & Phillips, 1998) that stands out in my memory. As an introduction to an investigation, students looked at graphs of linear relationships, one of which was of the equation $y = 300 + 2x$. With its graph on the overhead and pointing to the y-intercept, I asked the stu-

A Decade of Middle-School Mathematics Curriculum Implementation, pages 211–216

dents, "What is this?" I expected students to identify the *y*-intercept. Instead students said, "That is the starting or initial amount." I slowly thought to myself, "Yeah . . . yeah, that's right!" That was exactly right. Until that minute I had never made the connection between the graphical representation of the *y*-intercept and the situation it represented. In my mathematical experience, I had been taught to look at tables, graphs and equations. "Story problems" were rarely assigned. Even when I was asked to look at all representations, I was not pushed to understand the connections between them. For me, algebra had been about following a set of rules. It was not about understanding patterns of change and how those were reflected in other representations. Geometry was something that was taught separately, not as something that was integrated and connected to algebra. Now here I was witnessing students making sense of mathematics, and making the connections I lacked. Students were teaching me how different representations of functions were related, and it was exciting.

FIRST YEARS OF TEACHING WITH TRADITIONAL TEXTS AND *CMP*

My first year of teaching was in a district that did not use *CMP* and I did not enjoy the experience of teaching with a more traditional textbook. The text presented students with examples and they were expected to be able to repeat the procedures with practice problems. Before student teaching with *CMP*, I might have thought this was how mathematics should be taught. After all, it worked for me, why not for them? But after student teaching with *CMP*, I knew better. I knew there were materials that resulted in better student understanding of mathematics. Unfortunately, I was too young and inexperienced to try to get the *CMP* materials adopted by my school.

The following year a position opened up at Traverse City West Junior High, teaching 7th- and 8th-grade *CMP*. I was very interested in having another opportunity to work with this innovative curriculum. Even though my exposure to it while student teaching had been minimal, I knew that I wanted to teach with *CMP*, because it helps students understand mathematics, not just follow procedures. When I learned I had gotten the job, I was ecstatic. I knew I would have to work really hard, but the payoff would be worth it. I knew this was how mathematics was supposed to be learned. Working in Traverse City Area Public Schools also offered an opportunity for me to grow and be challenged professionally. The district was on the cutting edge of the teaching of mathematics and teaching practices in general and it offered many opportunities for professional development.

WORKING WITH A MATHEMATICS COACH

I soon discovered that teaching mathematics with *CMP* would involve much more than picking up the book the day before class and figuring out which homework problems to assign. Fortunately, there was an incredible amount of professional support for me. At the time I was hired, Yvonne Grant was funded by a grant to work as a mathematics coach with our district. From her I received one-on-one training for each unit I taught. As I became more comfortable with the mathematics, Yvonne began to pop into my classroom to function as a coach. Because I had a great amount of respect for her and had developed a rapport with her through the unit training, having her in my classroom was invaluable. Each time she came to observe, she focused on whatever area I wanted to improve, whether that was questioning, assessment, or anything else. After each observation, we would sit down together and talk about the lesson. The discussions were always insightful and nonjudgmental. She was pushing me to be a better teacher, but it was done in such a supportive way that I did not even realize it was happening. Of all of my experiences with professional development, I would list the time with Yvonne as the most beneficial.

Through my district I also became involved in the Michigan Algebra Initiative. This was a project run by Michigan State University for a few districts throughout the state. The objective was to look at student performance on tasks to determine what students knew about algebra. The study participants came from schools that used traditional curricula and from those that used problem-based curricula like *CMP*. One aspect of the project was to focus on student work. On one particular problem, students were asked to "describe the relationship." I was appalled by my own students' responses when I compared them to what other teachers had brought in. Their students had really done a nice job of being thorough and descriptive of the mathematics. My students had often written one word. I realized that maybe I was not expecting enough from my students in terms of explaining their thinking. From that point on my expectations began to change. I began to think more deeply about how I might use the structure of *CMP* lessons to help students develop quality explanations. For example, each investigation in a *CMP* unit ends with *Mathematical Reflections*. At the beginning of each school year I write out these mathematics reflections together with the students as a way of modeling what I expect. I make specific suggestions for things to consider when writing reflections, such as including examples to demonstrate their thinking. Later in the year when students are writing their individual reflections, I give them opportunities in class to talk about their mathematics reflections with one another. As a result of these small interventions, the reflections students now write are far more explicit and detailed than when I started 9 years ago. I use the reflections to learn what students do and do not understand. This helps me to make instructional decisions either for the individual or for the class as a whole.

CHANGES IN ASSESSMENT PRACTICES

When I first started teaching, my assessment was only focused on giving grades. I struggled with how many points to assign, how to give partial credit, and so on. I spent my first couple of years writing tests that would trick students. I did not look at assessment as a way to gain insight into what students knew and understood, but rather a way to find out how much they did not know. I can remember getting ready to give a Check-Up in the unit *Growing, Growing, Growing* (Lappan, Fey, Fitzgerald, Friel, & Phillips, 1998). I looked over the problems and thought it was not going to be a good quiz because every student would get 100%. I thought a quiz should challenge students and this one was obviously a piece of cake. It did not occur to me that every student getting 100% might show that the students had mastered the mathematical objectives and, of course, this was good!

My limited view of assessment began to change in 1998 when I became involved in a Balanced Assessment project. The participants in this project included nearly 100 mathematics and science teachers, district mathematics and science coordinators and curriculum specialists and educators in institutions of higher learning. The project director for mathematics was Sandra Wilcox. The aims of the project were to help teachers learn to assess for understanding and to use assessment information to reshape their teaching practices in order to make instruction more effective for all students. Through this project I began to grapple with my limited view of assessment. Instead of looking at assessment as simply a form of evaluation at the end of a section, I began to look at it as something that could inform my teaching. I gradually started to see assessment from a much broader perspective. I began to watch students much more closely during the "explore" phase of the lesson. I took notes about which students were doing what. I thought about how I could bring their work into the "summary" part of the lesson. I was assessing their understanding during each class period, not just at the end of the section and I used my observations to plan: to plan for the next day, the next week, and to individualize instruction.

CONTINUED LEARNING THROUGH AND ABOUT *CMP*

CMP has many problems that have high cognitive demand. I found it a challenge to understand what this means and to maintain this high level through each phase of the lesson. During the "launch" phase of a lesson, it is so easy to give a problem away by giving out too much of the mathematics. When this happens, the students are left with nothing to figure out on their own. During the "explore," when I see students struggling I am tempted to swoop in and save them by scaffolding away the struggle. The "summary" phase was the hardest part of the *CMP* structure for me to learn. It was hard to orchestrate the conversation in such a way that the students had a discussion among themselves, not a "ball bounce" with me. I had to learn how to pull out the mathematics in a way that it became transparent. Group work presented its own issues for the "summary" phase. If I had students do work on big poster paper, how did I decide who should present? How many presentations were necessary and in what particular order? Should I post someone's work that was incorrect? How did I make sure everyone understood? How did I hold all members of the group accountable?

I found it challenging to move from making sure students had the right answers to making sure they understood the mathematical objective of the problem. It took a lot of experience to know the problems, the connections between problems and connections to other units. Knowing where students had been mathematically and where they were going was very helpful. I write of mathematical understanding for the students, but I needed to have it as well. My deep understanding of the mathematics was critical for knowing what mathematics to pull out of each of the problems. The authors of *CMP* were very deliberate about the purpose and placement of particular problems and their purposes. It took time for me to fully understand these purposes.

As I became more comfortable with the mathematics, my questioning began to improve. While planning out a lesson, I would think about how students might approach the problems, what strategies they might use, the roadblocks they might run into, and the misconceptions they might have. How and when would I address these? What questions would I ask during the "explore" phase for students who finish early? What questions would I ask those students who were struggling? What would be important for me to watch for in the "explore" phase and what questions would I ask in the "summary" phase to be sure they had understood the mathematics? Much of this detail can be found in the teacher's edition of *CMP*, but it takes a lot of time to absorb all the information. I found it was important for me to be as specific as possible with all of this. When I was well prepared, students were able to walk away with a much deeper understanding of the mathematics and to see connections.

It took me a long time to see the progression of content across units and grade levels. In one of my first years with *CMP*, I can remember early on teaching students how to use the distributive property. We spent days and days on it. I did not realize that the concept would be addressed again in a later unit. The developers did not intend students to develop mastery at this first encounter, but rather a beginning understanding. As I became more familiar with each unit, the connections between them became clearer. I also became better at triggering their memories of the mathematics with questions about the contexts such as, "Do you remember the Wumps?"

WORDS TO THE *CMP* BEGINNER

I feel very fortunate to have been able to grow as a teacher with *CMP*. If asked to give advice to teachers new to *CMP* I would say to look to their colleagues. The incredible support I had from my professional peers has been most helpful. My colleagues were always available for conversations between classes and to talk about how lessons were working. I was able to co-plan with a fellow mathematics teacher on a weekly basis. I would recommend finding a colleague who teaches mathematics at the same grade level and setting aside time on a weekly basis to talk about what is going on in your classroom. I also recommend participating in as much professional development as possible. I had a strong desire to be successful and I was willing to put in the time and effort to accomplish this. I participated in any professional development that my district offered and I looked for opportunities to make workshop presentations. These experiences helped me think deeper and reflect more on how students learn and what is important for them to understand. My final advice is to have fun teaching with *CMP* and know you are doing your students the best service possible by providing them with a solid mathematics background that helps them become better at problem solving and communicating mathematically.

CHAPTER 22

A STORY OF TEACHER GROWTH

Teacher as Learner

Barbara Diliegghio
Waterford School District (retired)
Waterford, Michigan

> Consider this scenario: *You have been teaching middle grades mathematics and science for 24 years. The day comes when, although it still feels like you are doing the right thing (teaching), it just doesn't seem like you are doing it "right" anymore. You are working hard to prepare and deliver informative lessons. Your students listen as well as can be expected and they try everything you have to offer. How come they often just don't "get it," despite consistent effort on your part and theirs? You think to yourself that surely you can do better and you certainly need something better to offer your students.*

That was me about 8 years ago. Many teachers face similar turning points in their careers. Some of them quit teaching and some of them keep plodding along doing the same old thing. I decided to try to do a better job. This is the story of my efforts to do better and the resulting changes in me, my students, and my colleagues.

A Decade of Middle-School Mathematics Curriculum Implementation, pages 217–224

LOOKING FOR A BETTER WAY

My first approach to getting better at teaching was to seek out new ideas. While at a conference for science teachers, I learned that achievement is higher when science is taught in real-world contexts. I wondered if this was true with math, too. Would teaching mathematics through real-life contexts help my students learn better? I tried to search the internet for lessons that embedded mathematics in contexts. This task proved to be both time consuming and of limited value, as many of the projects I found were of low quality or more appropriate for elementary classrooms. Next, I tried to create my own materials, but I soon recognized how challenging that was. However, when I tried the few lessons I did find or create, I was encouraged. I realized that when my students worked on a mathematics concept that was applied to a real-world context, they seemed more interested and better able to understand the mathematics embedded in the context. This result energized me and so I kept looking for more materials.

About that time, one of the universities in my state was awarded a grant to work with teachers who were interested in reforming their mathematics classrooms. When information about the grant circulated in my school, I signed up. I was not sure what reforming mathematics classrooms meant, but I knew I was likely to learn something and I was not going to have to pay graduate tuition to learn it. During workshops sponsored by the grant, we reviewed findings of current research in mathematics education and we completed lessons from various NSF-funded curricula to see how the research findings applied. At the end of the grant I had to choose one of the middle grades programs to use in my classroom. I chose *Mathematics in Context* (*MiC*) (National Center for Research in Mathematical Sciences Education & Freudenthal Institute, 1997–1998; Wisconsin Center for Education Research & Freudenthal Institute, 2006) because I thought it was the most mathematically rich and the contexts I had been looking for were woven throughout the curriculum!

That was the beginning of my search for something better and, as with many things in life, it was a "good news–bad news" story. *MiC* was the good news. I will have more to say about that later. The bad news was that I was the only person in my school, in fact the only person in my district, using any NSF-funded program. I was in this *reform mathematics* boat all alone and paddling against the current! But, as they say, "ignorance is bliss," and I took the *MiC* materials and joyfully went back to school to start using them with my students.

I put away my old textbooks, and taught my first *MiC* unit in May. I must have been out of my mind to start something like that in May, but I am glad I did. Even after one unit I started to notice a change in my class. Those 6th-grade students started to focus more and they wanted to share what they were learning. They actually started talking about mathematics instead of what they were going to do after school. They allowed me to put them in groups (even boys with girls!). Everyone was working (remember this is May!) and I started to learn as well. I started to learn what my students were thinking and I could not believe that they were solving problems using a variety of strategies. This really started to shake my belief system, because I always thought there was only one way to do math. Boy, was I ever wrong!

I decided to use *MiC* exclusively the next year. This was a very bold decision fueled by my students' reactions to the first unit. Even though I did not have much proof, other than the data I collected with my students on their exams, their excitement about mathematics and the way they were transformed from passive learners to empowered thinkers convinced me that this was the right thing to do for my students.

CHANGES IN MY TEACHING

My biggest challenge in using *MiC* from a teaching point of view was to stop doing all the work for my students. I was accustomed to telling my students what to do and how to solve problems, and although it might have worked to a degree with a traditional program, it was clearly not appropriate for *MiC*. Eventually, I realized that learning mathematics was not about me, it was about the students. Getting my students to a higher level of thinking and deeper understanding required me to stop telling, and start asking.

I found some key general questions that I could use in almost every situation. "What do you think? How did you get that answer? What is your proof? Does this always work? How do you know?" Those questioning strategies were so new to me I almost had to write myself a script. At first, I had trouble watching the students work, identifying the disconnect in their thinking, and then creating a question to facilitate the students' thinking so they could continue to work on the problem. So when I prepared for a lesson, I imagined a "practice run" with my students, all the time thinking about what questions they might ask, what problems might come up in class, and how much prior knowledge and understanding they would need for the lesson.

That last question about prior knowledge made me realize that I had no idea what they really knew and that was scary. I would fall asleep pondering these questions, and when I woke up I would worry some more. It was exhausting, but I knew all my angst was worth it, because every day I saw a child experience deep understanding, I heard questions that were thoughtful conjectures, and I witnessed students sharing strategies that I had never considered. The celebration of discovery and learning was the food that fed my efforts and kept me going each week.

My level of job satisfaction changed because I went from "telling" all day to asking questions and facilitating. My old style of instruction required minimal thinking and preparation, but I was bored and I always went home tired. When I started to use *MiC*, I was constantly moving around the room, observing individual students, analyzing what I had observed, making instructional decisions, and then deciding on the question or strategy to address the needs I had observed. For the first time in my career I really knew what every child was thinking. I knew their strengths and what strategies each child used to make sense of the concepts. I found it amazing that I knew where every child was in the learning process as we progressed through the curriculum. I did not have to make assumptions or wait until the end of a unit to find out how much they understood. My personal level of activity and thinking were considerably higher, and much more satisfying. I still went home tired, but now it was a good tired, not a bored tired. As my students became more empowered to take charge of solving mathematical problems, I too became more confident of my abilities and efforts. I realized that I was not a failure if I felt like a new teacher even though I had been teaching for more than 20 years and that my students would not "break" if I made a mistake or they struggled.

Another revelation about learning that struck me as I traveled through the implementation process was that just as I discovered my students learn by "doing" math, and just telling them did not mean much; the same was true about the way I was learning. I needed to experience teaching the curriculum, running into issues, resolving concerns, and struggling to find better strategies. Learning is still an experiential process and I needed to welcome it, work with the process, and feel confident that change is a good thing. What was missing for me was being able to experience this learning with other teachers in my school through collaboration. I saw the value of my students talking about problems with each other, but this dimension of my learning was for the most part absent. What I did find, however, was that I was in a collaborative effort with my students. As their excitement and confidence to solve mathematics problems grew, so too did mine.

CHANGES IN STUDENTS

The changes I saw in myself were subtle and progressive over the first couple of years of using *MiC*. The changes in my students happened more quickly and were more dramatic. I saw passive learners turn into independent thinkers, and students who thought getting a wrong answer was a bad thing become risk takers who were willing to make and test conjectures. I saw students who had been "sitters" turn into problem solvers. They may not have been using the traditional algorithms, but they were still able to get the correct answer and explain the strategy they used. As I listened to my students, I heard discussions, debates, compliments, and questions instead of silence or whining. The class became a community of learners instead of solitary individuals working in their own little space. The whole energy level in the classroom rose and misbehavior diminished. Finally, the greatest change I witnessed was the change in attitude and respect for each other as students learned that everyone could be successful.

The decision to implement *MiC* had been a leap of faith for me and at first I lost sleep at night wondering if I was wrong. I had many questions and doubts: "Will they do well on the state exam? Do they really understand as much as I think they do? Was it the right choice to introduce them to a variety of ways to solve problems, or should I just stick with the standard algorithm? I chose to do what I think is best for kids, but am I cheating them? What if I am wrong? Did I teach a topic well enough? What is in the rest of the units? Will they cover the topics I am supposed to teach?" At the end of the year I had data and all the positive observations from my classroom to support my decision. My students did well on all the standardized tests. When I compared the scores of my students to other classes, we ranked the highest. Not only was there the usual small group of students that always did well on the exam, but now all of my students showed improvement and many were scoring at grade level or higher.

REACTIONS FROM PARENTS

When I first started using *MiC*, the parents of some of my students went on the attack with questions like: "What kind of mathematics is this? I never did this math. My child says you won't help him/her; why don't you just tell him/her how to do it? How come they are working in groups? Why do they have to write in math? Isn't it cheating if they don't do long division? I want my child to learn how to do it the same way I learned. Is this fuzzy math?" Those parents challenged me to be a good communicator and develop strategies to further understanding and collaboration. I developed a list of questions and strategies to help parents work with their children. I discovered the power of the mathematics journals, helping students to use them as a reference and share their work with their parents. I learned to be flexible, trying any strategy offered by a parent. I even had parents stay with their children after school so they could get help and learn the mathematics, too.

Other parents were my champions from the beginning with comments like: "I have never seen my son/daughter so interested in math. I love the way you help them to understand the concept as well as the strategies. The way my child can solve a problem is fascinating, I wish I had learned mathematics this way. Listening to my child has helped me understand math. You are very brave to try something new, I bet you are getting a lot of questions; but stick with it because this curriculum is amazing. I love the way my child can think about a problem using a variety of strategies. I want to thank you for working with my son/daughter in math; this is the first time that he/she has been successful."

Most of the parents were willing to trust me after we spent time conferencing about the program as well as the reasons behind my teaching strategies. I tried to build their confidence by proving I was knowledgeable and truly cared about their child's success. I must admit that not every parent "came on board." I remember one parent had her child transferred because she knew he was better than all the other students and didn't need to waste his time working with a group.

REACTIONS FROM COLLEAGUES

The hardest part of my decision to use *MiC* was that I was the only teacher in my building using it. It was often very difficult for a variety of reasons. First, I had to figure out the problems by myself and resolve issues without any collegial input. When I had questions about the curriculum, I emailed my questions to the authors as often as possible. I was relentless. I applaud their patience with my numerous questions and concerns. The only other source I had was the Teacher Resource and Implementation Guide (Cole, Romberg, de Lange, & Foster, 1998) and the teacher guides for the units. These resources became my friends. I would read them, reread, digest the information, and then reread them again. I would work through the problems, read every word in the support material, and then write my lesson plans.

Another way I felt alone was when my students started to problem solve, enjoy the discussions, and try new strategies. I did not have anyone to share my excitement with when I had mathematical "ah-has." But, more than that, it was very difficult to put up with the comments from other teachers who were convinced I was having too much fun to actually be doing a good job at teaching math. We were a traditional junior high where "good teachers" kept students quiet and busy. Some of the teachers would frown as they went past my classroom. My students were all talking to each other in groups and having fun. My colleagues started to wonder what was going on. I could almost hear their questions: What is she doing in that room? Why weren't the students in rows? Why are they so noisy? Has she lost control of her class? It couldn't possibly be mathematics if students were laughing and having fun. One teacher actually said to me, "In this school, we keep them quiet and in rows." I did not know how to react. I wanted to be accepted by my peers . . . I wanted to fit in. Yet, I saw the impact my instruction using the *MiC* materials was having on my students. I asked myself, "Do I do what is easiest for me or what is best for my students?"

The more I tried to explain what I was doing and invited my colleagues to join the adventure, the more I seemed to threaten their world. They were used to the comfort of "business as usual" and seemed to ignore the low test scores and the inability of their students to "use" math, often saying, "What can you expect with the kids we have in our school?" I wondered if they were afraid to learn new things. I found myself unable to answer the biggest question, "If schools are institutions dedicated to learning, why are teachers afraid to learn?"

Once I made the decision to continue using *MiC* and accepted that this was not a popular decision, I began to work at understanding why my peers were reacting in this fashion. I took a class on organizational behavior, and learned that change in an organization can result in the behaviors I had observed. I also learned that although there are some people who feel uncomfortable when the "status quo" is threatened, there are others who embrace change. I decided to stop feeling defensive and start looking for those colleagues, and to my delight I found them. They had always been there, quietly watching what was happening. So, the next year, we started to work together and I was not as solitary in my efforts because some of them were willing to try a unit. I found myself in a "coaching" role. More importantly I was collaborating: sharing student "light bulb" moments, but also problem solving when lessons did not go as planned.

I changed as a teacher through this implementation process, but I also changed personally. As our band of *MiC* pioneers slowly grew each year, I discovered through my role as a coach that I really liked working with other teachers. I decided to start presenting at our local and state conferences, and discovered this too was fun. Now when I do professional development with teachers and they seem frustrated, I try to figure out what they are experiencing. I think to myself, "Do they feel uncomfortable with the math? Is it the curriculum design? Do they think that they are the only one struggling with something new?" I try to help them realize change is a process and that we learn by "doing." I try to offer the support and collaboration I did not have when I started on my journey.

FINAL THOUGHTS

Looking back, I grew so much personally and professionally. I realized I could be the poster child for the "Old Dogs Can Learn New Tricks Society." I discovered a new richness in my life when I took on the challenge of change and learned new ways to do things. I learned that life and teaching is a process and that I am capable of problem solving, thinking, and communicating. I discovered that teaching is a craft and it must be nurtured and developed. I am amazed that even after 28 years, teaching—working with children as they learn mathematics—is never the same. I learned that facilitating children's learning of mathematics can be scary and exciting, but it should always be about doing what is best for the kids. So now I enjoy the process. I jump in, welcome what comes next, and look forward to what lies ahead.

CHAPTER 23

A STORY OF TEACHER GROWTH

Making a Difference

Kelly Watts
Regional Education Service Agency II, Kenova, WV

When I was in my late twenties, I made the huge decision to change careers and become a teacher. I had been educated as an accountant and worked in a bank for several years, but something was lacking... I wanted to really make a difference and had toyed with the idea of becoming a teacher for some time. Simply put, I believed that I could make a difference in students' lives. Like many beginning teachers, I had all sorts of idealistic reasons for wanting to teach. I approached my first years in the classroom with a lot of energy, hard work, and zeal. I showed students how to do mathematics in the day's lesson and then I had them practice the procedures in class and do more for homework. I regularly checked homework, gave quizzes and end of chapter tests in order to assign grades. I did great explanations and kept students quiet and in straight rows. I occasionally used group work—every 6 weeks or so—when I would try to do some form of fun activity. I was doing everything I knew how to do and more. I often gave up my planning period and lunch to help students with homework and I even received a Title I commendation for this practice. Had you observed my classroom during those first five years you would have seen a pretty typical class... I was teaching the way I had been taught.

A Decade of Middle-School Mathematics Curriculum Implementation, pages 225–230

DIMINISHED HOPES

Unfortunately, I did not get the results for which I had hoped. My efforts seemed to have little impact on the students and I knew they were not really learning mathematics. They made few connections between topics and because of low retention I felt I constantly had to re-teach what I thought they should already know. They were doing the least amount of mathematics they needed to get by. Worst of all, many students were "giving up." Nothing I did seemed to have any effect on their attitudes and their learning. With little or no sustained success I was becoming very frustrated. I was not making the difference in students' lives that had brought me to teaching. I needed a better way of teaching and engaging my students.

Early on, being the lifelong learner that I am, I continued to seek professional development opportunities and I was fortunate to have several available. I began by attending summer professional development in mathematics. I would get a lot of ideas for activities that I could use in my mathematics classes, but when I returned to school in the fall I would feel overwhelmed. Sometimes the activities would work, but often they would fail and discipline was a problem with many of the interactive activities. Collaborative learning did not work and I realized later it was because I was expecting students to collaborate on problems that did not require collaboration. To make things worse, I was the only one at my school attending this professional development and only one of three teachers from my county. That left me feeling pretty isolated and without resources for making needed improvements.

PROJECT MERIT

A significant turning point in my teaching career came when I applied and was accepted to participate in a statewide National Science Foundation project called Project MERIT (Chapter 7 by Murrel Hoover fully describes Project MERIT). The Project was for middle level teachers like me and we were to receive some new materials to use in our classrooms. This is where I first encountered the *MathScape* (Education Development Center, Inc., 1998, 2005) curriculum and during this first year I was given intensive professional development on three *MathScape* units with the understanding that I would use them during the next school year with at least one class.

During the summer sessions, Amy Doherty, an experienced *MathScape* teacher from Education Development Center (EDC), provided the intensive training. I found this experience to be one of the best professional development experiences I would ever have. Amy arranged us in collaborative groups to work together as students would work on the mathematics in the unit. Working with other teachers was great. We supported each other's learning and I began to see how powerful learning mathematics in groups could be. We eventually (over a two-year period) worked through six of the seven curriculum units in this manner. Additionally, the project provided professional development that supported our implementation in other ways such as writing in mathematics, collaborative learning, learning and teaching styles, understanding the NCTM Standards, assessment, and so on.

In addition, there were follow-up experiences on four Saturdays each year with a focus on pedagogy. We read books and articles and participated in book studies and small group discussions. Through these conversations, we became more aware of how students learn mathematics, and we learned to study student work to better understand our students' thinking. Sometimes we divided into groups by grade levels and all the teachers at a particular grade worked together and shared expertise. At other times, we worked in cross-grade groups to explore how the mathematical ideas grew across the grades.

For me, the most support from the project was from participation in WebCT. In this web-based forum, I could connect with other *MathScape* teachers across the state 24–7. These grade-level specific and general web-based chats really contributed to my growth. They became a forum in which I could ask questions and receive suggestions from colleagues. I work a lot at night, so sometimes I would post a question or a success story and get several replies throughout the evening and into the early morning hours as other teachers sought the same collaboration. We began to collaborate in ways we had never considered...we created and posted resources to complement each of the *MathScape* units. Even if we were not teaching the exact same unit at the same time, the resources and discussions stayed there and I could refer to them whenever I was teaching that unit. Resources included alignments between the state standards and the curriculum, pacing guides, and additional charts or grids or graphs that teachers designed to help teach the unit. These shared resources were excellent and a great time saver. The collaboration was wonderful!

MATHSCAPE IN MY CLASSROOM—RENEWED HOPE

I wish I could say that all the professional development training I received prior to Project MERIT made it easy to bring *MathScape* into my classroom, but it didn't. The first year was tough. However, as I taught the units I realized that the professional development did help me to be successful. During my first year teaching three of the *MathScape* units, I found myself getting hung up on the issue of basic skills. I was not sure students could do some of the mathematics in the units and I ended up spending too much time on reviewing basic skills and trying to pre-teach the mathematics the students were to explore. I simply did not trust the curriculum. Thankfully, through the web-based discussions with other Project MERIT participants, I found other teachers with promising results, who were not worried about basic skills. I made a promise to myself to relinquish some of the tight control I had on the class in order to give students more room to explore and become more independent.

Besides the basic skills issue, the next biggest challenge for me was the time needed for planning and management of the curriculum. Unlike the traditional materials I had been using, I found that I had to have a clear plan for the unit I was teaching... a plan for the *whole* unit before I began to teach the first lesson. I devised a notebook that would help me in subsequent years. In this notebook were the handouts I needed to copy, materials lists, suggested additional skill sheets, and, most importantly, my reflections on how the lesson had gone when working with students. I also had to have a clear assessment plan. *MathScape* offers in-depth assessment opportunities and I had to structure my lessons to make the best use of them. I found that I really started to learn a lot about what my students knew from their *Dr. Math* letters. (*Dr. Math* letters were writing prompts about the mathematics in the lesson.) At first, it took me forever to grade these, so I designed a rubric that allowed me to focus on what I needed to assess. This saved me a lot of time and also afforded me a clearer picture of what was important to assess.

Everything I did with *MathScape* seemed to take more time. Planning took a lot more time than I had anticipated, as did gathering materials I would need in my lessons. Because of class time constraints I had to learn how to pace whole-class student discussions—to make decisions to keep students on course. I had to decide how often to keep groups together, or when to work in pairs or individually. While making all these decisions, I still had to figure out if and what my students were learning. The writing piece helped me to identify what the students understood and forced me to design more efficient ways to recognize students' strengths and weaknesses.

POSITIVE CHANGE

I wish I could say that I fully embraced the ideas of the *MathScape* curriculum at the onset, but that would not be true. Despite the frustrations with my previous curriculum, I was initially somewhat reluctant to make the changes necessary to teach *MathScape*, but I began to see significant changes in my classroom and in my students that gave me hope that all this effort was worth it.

One of the first encouraging signs was that my classroom had fewer discipline issues than before I started implementing *MathScape*. As I loosened control, the students began to take more ownership of their mathematics learning. They began to talk to each other about mathematics, even outside of class. I saw students helping other students and that had never happened before in my class. Students started to make connections and they would say things like, "Now I know why I need that!" Mathematics became more accessible and more usable for them. For the first time, I saw students enjoying the learning of mathematics. I saw that the curriculum addressed the variety of learning styles my students brought to mathematics class. I saw *more* success for *more* students *more* of the time. When I saw my students experiencing success, I realized I was starting to reach the goal of making a difference in mathematics for kids.

I also saw big changes in my teaching. My pattern of show how, practice, and assign homework again and again, gradually vanished as I began to trust the curriculum. I now go over homework only as it leads to the day's lesson. Instead of showing how to do the mathematics, I pose a situation or problem and facilitate discussion, thus allowing the students to be more engaged in their learning. I still assign homework but it is more focused on the objectives. For example, it might just be a journal entry instead of working on many problems. Journal entries reveal how students think about the mathematics and that enables me to arrange the group work to facilitate discoveries in class. I now actually *know* what my students know or don't know.

I also found that I was learning new mathematics and deepening my understanding of the mathematics I thought I knew. For example, recursive patterns started to make sense to me instead of just being a procedure to follow. I also learned to visualize three-dimensional geometry. I had always struggled to *see* things. I could do the problems by procedure, but I never could *see* it in my head. I also have a deeper understanding of some of the big ideas like proportional reasoning and multiple representations and I can see how these connect to algebra and geometry.

PARENTS

I want to end with a story about parents. When I began to use *MathScape* the parents were either frustrated or excited. The frustrated parents were the ones whose children did well in elementary school but started having trouble in middle school and the parent could no longer help. The excited parents were the ones who liked seeing their children challenged, for maybe the first time, in a positive way that promoted good mathematics.

I remember one parent in particular whose daughter (I will call her Elizabeth) was in my class for three years. In grade 6, Elizabeth made her first

C in my mathematics class and the mother was upset. However, this grade was appropriate because Elizabeth had not yet learned to think critically or solve various problems until I directed her thinking. She was struggling, but in a good way. In grade 7, Elizabeth began to show more confidence and attempted to work out problems before she came to me for help. She began talking more with her classmates about strategies and she realized that there is usually more than one way to solve problems. However, since Elizabeth was still making Cs and Bs, her mother was frustrated and she came to tell me she was requesting Elizabeth be placed with another teacher for mathematics in grade 8. The mother reported that mathematics was causing a lot of stress in their family. In contrast, Elizabeth seemed happy in class and she did not want to switch teachers because now she recognized that she was learning. With a lot of reluctance the parents agreed for Elizabeth to continue with me the next year. In eighth grade everything seemed to come together for Elizabeth and she started to make Bs and As. The real end of the story came a year later—after Elizabeth had completed Algebra I in high school. The mother visited me (I just hoped she wasn't coming to complain). She thanked me for all that I had done to help Elizabeth become successful in mathematics. Those were nice words to hear, but it was not so much me who made the difference, but rather the curriculum I had to support me. *MathScape* made a difference for me as a teacher and that helped me make a difference for many students like Elizabeth.

CHAPTER 24

A STORY OF TEACHER GROWTH

From Telling to Facilitating

Fabienne Anderson
Ariel Community Academy, Chicago, Illinois

I began teaching middle grades mathematics in 2002. I taught in a school that specialized in using rich curricula to educate at-risk students. The school, although small, was a part of a large, urban school system. When I arrived I found on the shelves a traditional text that had been ordered by the previous teacher. I was given the choice of using that curriculum or ordering another. Although I was new to the middle grades, I was not new to teaching. I had taught for 10 years at other grade levels and I was very familiar with traditional texts. It had been my experience that I often had to supplement and add to those materials in order to broaden my students' understanding of the mathematics concepts presented. I went into the new middle school position with the understanding that I could choose the curriculum I felt most comfortable teaching and MATH*Thematics* (Billstein & Williamson, 1999a) was, and still is my choice. I realized that not all teachers have this same privilege, so I took my time in deciding on the right curriculum. I decided to use MATH*Thematics* instead of the more traditional

A Decade of Middle-School Mathematics Curriculum Implementation, pages 231–234

curriculum, because I felt it was a better match to both my teaching style and my students' ways of learning.

I feel it is important to allow students to discover concepts and to explore the relationships between mathematics concepts with real-life phenomena. I find that because my students learn through discovery, retention of knowledge is often greater and easier for them to build upon what they already know. For example, my students remember the concept of circumference because of measuring various lids and comparing the diameter to the distance around the lids. Because of that hands-on activity, my students remember that the ratio of the circumference to the diameter is a constant we call pi.

After choosing MATH*Thematics*, I embraced the curriculum and eagerly approached teaching the modules. I felt such excitement when exploring and planning the lessons and I displayed that same enthusiasm when I taught my students. One of my favorite modules, *Mystery of Blacktail Canyon* (Billstein & Williamson, 1999a), is based on a make-believe mystery. I enjoy teaching that particular module because of the way the students approach the mystery. They really appreciate the mathematics involved in identifying suspects and solving some of the mini-mysteries within the sections. Because I demonstrate excitement when exploring mathematics concepts, my students begin to embrace and value the mathematics too and to experience higher mathematics achievement.

After I chose MATH*Thematics*, I was offered monthly professional development for almost a year to make my transition as easy as possible. I looked forward to meeting with "my" publisher representatives as they seemed to share my enthusiasm and zeal in presenting the explorations. I brought to the sessions my questions and successes and the representatives shared their expertise on the curriculum and all its components. Now, since my second year teaching with MATH*Thematics*, I too serve as a teacher consultant and provide professional development to teachers who are new to the curriculum. I share with them my experiences as well as new ideas for teaching it.

One of the things I tell teachers in professional development workshops is the importance of working through the material before presenting it to children. I remember one of the greatest challenges I faced when I began teaching with MATH*Thematics* was discovering the concepts as the authors intended. I remembered how I had been taught mathematics and I remembered I had not understood a lot of it until much later in my educational career. I wanted my students to understand the mathematics and to see its value as they explored. This was possible with MATH*Thematics* because it allowed my students to study concepts in the context of real-life phenomena. When I began to teach with the curriculum, I failed to look at all the components and I was not prepared. I thought I could treat it like a traditional text and glance over it and be ready to teach with it. When my students got into the contexts and the mathematics, they had questions and I could only answer from my limited understanding. I had robbed my students of developing their own rich understanding because I had not prepared. I resolved to prepare for each lesson by working through each problem as if I were a student. I had learned my lesson well, and after that mistake things started to turn around for my students.

The first year of teaching MATH*Thematics* was the toughest because each lesson was new and I had to prepare many manipulatives to enhance my students' experiences. I cut out and laminated many manipulatives to save time. The second year was much easier because the material was familiar and I had saved everything in folders and plastic bags in anticipation of the next time they would be needed.

I saw many changes in me and in my students. I saw myself change from a somewhat didactic teacher to a facilitative teacher. The explorations allowed my students to discover concepts for themselves; as a result they took more responsibility for their own learning. No longer was I the only source of information and understanding. I was excited to see that I was learning new ways of looking at mathematics as my students worked through the daily explorations. For instance, I had never really understood box-and-whisker plots until I taught an exploration about these plots. I learned along with my students the use of that graph in communicating information about data. The spiraling of the curriculum allowed my students to use and apply box-and-whisker graphing to problems that were relevant and interesting to them.

There was a lot of graphing and technology use that interested my students and me. I had never used a graphing calculator until I tried some of the activities suggested in the Teacher Resource Books (Billstein & Williamson, 1999b). My students were amazed at what graphing calculators could perform, and after a very short time, my students had convinced their parents and members of our local school council of the calculators' importance. I received a class set of graphing calculators during my first year with MATH*Thematics*, and several parents purchased graphing calculators for their children to use at home.

Almost immediately, the parents of my students appreciated the impact MATH*Thematics* had on their children. Students' attitudes regarding mathematics changed quickly from negative to positive. They began to see the value of mathematics and their parents let me know. Several parents would tell me regularly, "He might not do anyone else's homework, but each night he says he has math to do." As my students began to take standardized tests, they experienced increasing mathematics scores as well as a sense of accomplishment. Students began to tell me they felt more comfortable about completing the mathematics portions of tests (grade-level, and in some cases, high school entrance exams) than any other portion of the tests. Parents described seeing the changes in confidence in their children and they were secure in knowing their children were properly prepared for secondary school with mathematics.

In conclusion, I would continue to tell teachers who were just getting started with a standards-based middle grades curriculum such as MATH*Thematics*, to take heed. It works! The time I have put into the preparation and presentation of the curriculum has paid off seven-fold. My students over the years have blossomed and become mathematicians who could compete with students all throughout the school system. Their standardized test scores increased steadily each year and, more importantly, their attitudes towards mathematics grew more positive with each module of each book. I watched my students' sense of mathematics efficacy increase and my students saw mathematics as a positive part of their futures. I would encourage teachers new to standards-based middle grades curricula to prepare, be tenacious, and watch in wonder as their students reap the benefits of a rich and meaningful educational experience. I did and continue to do so.

CHAPTER 25

RESEARCH MATHEMATICIAN AND MATHEMATICS EDUCATOR

A Foot in Both Worlds

Ira J. Papick
University of Missouri, Columbia, Missouri

I am a research mathematician and a university mathematics educator who has passion for teaching teachers. My journey to this point has been filled with opportunities to think deeply about mathematical content, and learning and teaching mathematics. This chapter is about how I came to identify myself (with a foot in both worlds of research mathematician and mathematics educator), how being in both worlds has impacted my views about learning and teaching of mathematics, and ultimately how I collaborated with others to develop a university-level mathematics curriculum for middle grade teachers.

A Decade of Middle-School Mathematics Curriculum Implementation, pages 235–244

MY ENTRANCE INTO THE WORLDS OF MATHEMATICS AND MATHEMATICS EDUCATION

I am not sure if the sweat on my brow was due to a state of high anxiety or the sweltering heat of a 1974, summer afternoon in Lincoln, Nebraska. The lecture hall was near capacity, and more than 100 researchers in the field of Commutative Algebra were eagerly awaiting the next scheduled talk. Several of the most distinguished minds seemed to be situated in the rows closest to the podium. A clock on the wall indicated 2:54 P.M., which meant that in six minutes the public launch of my professional career would take place. "Ladies and gentlemen, our next speaker is Ira J. Papick from Rutgers University, and his talk is entitled *Simple Overrings*." The introductory applause muffled my tympanic heart sounds, and within seconds, mathematical phrases were filling the room. It seemed to me that my style was graceful and smooth—that is, until I noticed my right foot had inadvertently found its way to the bottom of a waste paper basket. With the unruly can in tow, I briefly continued my spirited presentation, and then, suddenly, without notice, my foot found freedom. A huge roar of laughter combined with a massive ovation followed my "feat," and I managed to complete my talk without further acrobatics.

Immediately following the lecture, a scholarly-looking gentleman approached the stage and extended his hand for the traditional congratulatory greeting. This was not any ordinary mortal, but rather the father of modern Commutative Algebra, Irving Kaplansky of The University of Chicago. His thoughtful words to me were, "I found your results to be quite interesting, your enthusiastic teaching style most refreshing, and your coordination very impressive." Overwhelming excitement captured my emotions, and I knew that learning, creating, and sharing mathematics defined my future.

Unlike many of my colleagues, I did not fit the "math whiz" image as a child. Manipulation of numbers, symbols, and formulas was usually not a problem for me, but I was missing a firm understanding of the general principles and applications. I often remember asking natural questions like: "Why do we need to learn (know) this?" and "What does this have to do with everyday life?" My teachers' patent answers were: "You'll need to know this for the next test" or the more ominous remark, "You'll need this for junior high (senior high, college, occupation, life)." For many students this was sufficient justification, and for lack of alternative explanations, I reluctantly accepted the well-intentioned responses.

By the time I entered high school, my interest for mathematics was quickly waning. This greatly concerned my parents, and they arranged for me to see a tutor. As it turned out, Mr. Flood was much more than a mathematical coach. He helped me see the power and utility of mathematics and actively engaged me in the learning process. His methods of inquiry and exploration were foundational to my present instructional tenets and strategies, and my own personal struggles with algebra and geometry equipped me with an important perspective on "the learning" of mathematics.

My future was now clearly defined! I would become a high school mathematics teacher and spend my life spreading the mathematical word. Bachelor's and Master's degrees in Mathematics Education were to follow (six years later) and permanent teaching certificate papers (for New York State) were finally in my possession. I had worked very hard to arrive at this point, but something was missing. Although my undergraduate and graduate education provided me with sound mathematical footing and pedagogical practices, I felt that without a deeper and broader understanding of advanced mathematics, my ability to lead students on fruitful mathematical explorations would be somewhat limited.

I felt that a career modification was necessary. I would earn a doctorate degree in mathematics and devote my life to scholarship and the teaching of mathematics at the university level. Exciting, new instructional opportunities, involving both undergraduate and graduate students, would challenge me daily. Although my enthusiasm and passion for the subject would no longer directly impact middle and secondary students, the mathematical preparation of their teachers would be central to my collegiate life.

MY BELIEFS ABOUT LEARNING AND TEACHING MATHEMATICS

Three decades and more have now passed since that afternoon initiation in Lincoln. Rich classroom experiences and successful professional collaborations have thankfully borrowed the time. Practice and reflection have shaped and integrated my objectives and methodologies. As a result, three important ideas guide my thinking about learning and teaching mathematics: 1) engaging students; 2) inquiry for learning; and 3) mathematics for all. I expand on each of these ideas below.

Engaging Students

Learning is not a spectator sport! I believe that active involvement is essential—with shared responsibilities for teacher and student. The classroom should be a site for dynamic exposition and productive interaction. Supporting collaborative investigation and in-depth projects can combine with traditional lecture to form a strong pedagogical base. Technology (computer and/or graphing calculators) as a tool supporting exploration and computation is indispensable; and focusing on applications (both mathematical and real-world applications), motivates and stimulates interest, thus allowing the power and grace of mathematics to unfold through its utility.

Inquiry for Learning

Science in general, and mathematics in particular, evolve through a sequence of stages including (not necessarily in this order), observation, experimentation, exploration, analysis, conjecture, formulation, proof, and generalization. Mathematicians seek to understand fundamental structures by scrutinizing specific classes of examples and collections of data. The search for clarity and veracity is sustained through insightful questions and conjectures in unison with creative thinking and problem-solving strategies, and it is especially crucial for our students to comprehend and embrace these habits of discovery. Research institutions, with their experienced practitioners and vast resources, provide the vital environment to achieve these objectives.

Mathematics for All

The ubiquity of mathematics is evident, and its place in education is axiomatic. As educators of mathematics, we seek to empower <u>all</u> students with fundamental mathematical skills, which include number sense, computation (both written and technology-driven), problem-solving experiences, and a general awareness of mathematical diversity. Proper courses of study and subsequent student success are often dependent upon knowledgeable advising and dedicated mentoring. High expectations of student performance should be clearly communicated, and contrasting talents and learning styles should be respected and effectively exploited.

"Why do we need to know this?"

During my career as a research mathematician and as a teacher of mathematics, I have continually enjoyed the thrill of discovering (or re-discovering) mathematics and the privilege of discussing the beauty and utility of mathematics. Stimulating conversations with colleagues and students make each day exciting and unique.

My teaching duties have included a myriad of mathematics courses involving a variety of majors, many of which have been courses for preservice middle or secondary mathematics teachers. Since my research area is in

Commutative Algebra, I have special interest in Linear and Abstract Algebra for preservice teachers, and have developed and taught such courses over the years.

Although these courses were rich in mathematical ideas, the connections to important concepts in school mathematics were not always explicitly detailed. Several factors contributed to this shortcoming, but a most significant one was the lack of excellent curricular materials (both at the school and university level) to help illustrate and demonstrate the critical links. A natural consequence of this predicament was frustrated students, who asked questions like, "What does this have to do with teaching mathematics in middle or secondary school?" and bewildered professors, who asked "Why can't these students appreciate the beautiful theorems we are proving?"

And then there was light! The *Missouri Middle Mathematics Project* (*M^3 Project,* 1995–99), a National Science Foundation teacher enhancement project, provided an excellent solution to my preservice teaching dilemma. This project created a rare opportunity for mathematicians, mathematics educators, and middle grade mathematics teachers to collaboratively investigate exemplary NSF-funded middle school mathematics curricula. These innovative materials were replete with significant mathematical ideas (including some important topics that have traditionally been reserved for the secondary level), and were ideally suited to stimulate and reinforce fundamental concepts in preservice content classes. Future middle grade teachers would finally see vital connections between the mathematics they were learning at the college level and the mathematics they would someday be teaching to their students. The infamous question, "Why do we need to know this?" would no longer be sung in unison.

Mathematical Knowledge Needed for Teaching

A central point that emerged from our collaborative examination of NSF-funded curricula on the *M^3 Project* and later with the *Show-Me Projects I and II,* 1997–2007, is that the mathematical preparation of middle grade teachers needs to be revised and strengthened in order to successfully implement these new curricular materials. Teachers uniformly expressed that their college mathematics courses did not connect the mathematics they were learning with the school mathematics that they would be teaching. This message was certainly not surprising to mathematics educators, but it was to many mathematicians who had been teaching courses for future teachers. In fact, a common belief held amongst such mathematicians was that one could improve teacher qualifications by simply requiring prospective teachers to take more college level mathematics courses. Although this conviction was contrary to current research (Hill, Rowan, & Ball, 2005), it was (and still remains to some extent) a substantial obstruction to teacher preparation curricular reform at the university level.

To help stimulate and guide the mathematics preparation of K–12 teachers, the Conference Board of Mathematical Sciences (CBMS), in concert with the Mathematical Association of America (with funding provided by the United States Department of Education), developed the *Mathematical Education of Teachers Report* (*MET*) (CBMS, 2001). This document carefully lays out a framework to guide the development of mathematics content courses for prospective teachers. The framework built upon the premise that "The mathematical knowledge needed for teaching is quite different from that required by college students pursuing other mathematics-related professions. Prospective teachers need a solid understanding of mathematics so that they can teach it as a coherent, reasoned activity and communicate its elegance and power" (CBMS, 2001, p. xi). The report asks mathematicians to rethink courses for prospective teachers and provides the following curriculum and instruction recommendations for the mathematical preparation of K–12 teachers:

1. Prospective teachers need mathematics courses that develop a deep understanding of the mathematics they will teach.
2. Although the quality of mathematical preparation is more important than the quantity, the following amount of mathematics coursework for prospective teachers is recommended.
    a. Prospective elementary grade teachers should be required to take at least 9 semester hours on fundamental ideas of elementary school mathematics.
    b. Prospective middle grade teachers of mathematics should be required to take at least 21 semester hours of mathematics, that includes at least 12 semester hours on fundamental ideas of school mathematics appropriate for middle grade teachers.
    c. Prospective high school teachers of mathematics should be required to complete the equivalent of an undergraduate major in mathematics that includes a 6-hour capstone course connecting their college mathematics courses with high school mathematics.
3. Courses on fundamental ideas of school mathematics should focus on a thorough development of basic mathematical ideas. Along with building mathematical knowledge, mathematics courses for prospective teachers should develop habits of mind of a mathematical thinker and demonstrate flexible, interactive styles of teaching.
4. Teacher education must be recognized as an important part of mathematics departments mission at institutions that educate teachers.

> More mathematicians should consider becoming deeply involved in K–12 mathematics education. (CBMS, 2001, pp. 8–9)

These recommendations, combined with my evolving beliefs about mathematics teaching/ learning and my experiences with the funded projects described above, prompted a thought—what we needed in mathematics education was a set of university-level mathematics textbooks designed specifically for the education of preservice middle school mathematics teachers. In the remainder of this chapter, I present the project that brought this idea to fruition.

DEVELOPING A UNIVERSITY-LEVEL CORE MATHEMATICS CURRICULUM FOR MIDDLE GRADE TEACHERS

Using the *MET Report* as a basic framework, a group of mathematicians and mathematics educators at the University of Missouri–Columbia jointly developed four fundamental college level mathematics text books as part of the NSF funded project, *Connecting Middle School and College Mathematics* [$(CM)^2$], 2001–2006. The purpose of the design of these materials was twofold. First, they were designed to provide middle grade mathematics teachers with a strong mathematical foundation. Second, they were designed to specifically connect the university-level mathematics the preservice teachers were learning with the mathematics they would be teaching. The books focus on algebraic and geometric structures, data analysis and probability, and mathematics of change, and extensively employ content from the NSF-funded middle grade mathematics curricular materials in the following multiple ways:

- As a springboard to college level mathematics;
- To expose future (or present) teachers to current middle grade curricular materials;
- To provide strong motivation to learn more and deeper mathematics;
- To support curriculum dissection, that is, to critically analyze middle school content and develop improved middle grade lessons through a lesson study approach;
- To use college content to gain new perspectives on middle grade content and vice versa; and
- To apply middle grade instructional strategies and multiple forms of assessment to the college classroom.

The $(CM)^2$ materials were extensively piloted in summer institutes, in courses offered at school-based sites, through a variety of professional development programs, and in both undergraduate and graduate semester courses offered at a number of universities across the U.S. The mathematical content of the materials was significantly shaped by the thoughtful and insightful comments of those piloting the materials. There is no doubt that the resulting four books were greatly improved as a result of the shared expertise of the classroom teachers, mathematics educators, and mathematicians who closely examined the materials.

The $(CM)^2$ college level mathematics texts are now in print (published by Prentice Hall, 2005, 2006). The following brief descriptions provide an overview of the content of each book (for a more detailed description of the textbooks, go to: http://vig.prenhall.com/ and search for "Connections in Mathematics Courses for Teachers"):

- *Algebra Connections* by Ira Papick (2006). Topics include: a detailed study of the integers with special emphasis on the division algorithm, least common multiple, greatest common divisor, Euclid's algorithm, prime numbers, The Fundamental Theorem of Arithmetic, Pythagorean triples, modular arithmetic and algebra, rational and irrational numbers, and linear systems of equations and matrices.
- *Calculus Connections* by Asma Harcharras and Dorina Mitrea (2006). Topics include: concepts of calculus such as sequences, series, functions, limits, continuity, differentiation, antidifferentiation, and other mathematics of change connections to the middle grades curricula.
- *Geometry Connections* by John K. Beem (2005). Topics include: logic, similarity, surface area, volume, coordinate geometry, vectors, transformations, and the use of matrices in geometry. Some comparisons of Euclidean geometry with spherical geometry and taxicab geometry are included.
- *Probability and Data Analysis Connections* by Debra and Michael Perkowski (2006). Topics include: collecting and displaying data, describing data with numbers, correlation with regression, probability, methods for counting outcomes, random variables and probability distributions, estimation, and hypothesis testing.

Throughout each book, the reader will find a number of Classroom Connections, Classroom Discussions, and Classroom Problems. These instructional components are designed to deepen the connections between the mathematics that preservice teachers are studying and the mathematics they will be teaching. The Classroom Connections are middle grade investigations (taken, with publisher permission, from the four NSF-funded middle school mathematics curricula) that serve as launch pads to the college-level Classroom Discussions, Classroom Problems, and other related collegiate mathematics. The Classroom Discussions are intended to be detailed mathematical conversations between college teacher and preservice middle grade teachers, and are used to introduce and explore a variety of important concepts during class periods. The Classroom Problems are a collection of problems with complete or partially complete solutions and are meant to illustrate and engage preservice teachers in various problem solving techniques and strategies. The continual process of connecting what they are learning in the college classroom to what they will be teaching in their own classroom provides preservice teachers with real motivation to strengthen their mathematical content knowledge.

To further illustrate the nature of the $(CM)^2$ materials, I will briefly describe how an *Algebra Connections* lesson unfolds in a classroom of preservice middle grade mathematics teachers.

A Classroom Example

Typically the class period begins by dividing the preservice teachers (PSTs) into small working groups (2 to 4 per group). The PSTs then proceed to investigate some important ideas in mathematics as they are laid out in a specific lesson from a NSF-funded middle school curriculum. A unique feature of the *Algebra Connections* book (as well as the other books in the *Connections* series) is that it contains the relevant pages from the middle school materials needed for the intended explorations. For example, the class might examine a foundational lesson from the sixth grade unit in *Connected Mathematics* (Lappan, Fey, Fitzgerald, Friel, & Phillips, 1998) called *Prime Time.* This *CMP* unit focuses on prime numbers, factorization, least common multiple (LCM), greatest common divisor (GCD), and sets the stage for an in-depth college level study of the Fundamental Theorem of Arithmetic.

A particular lesson from *Prime Time* (Investigation 4.2, Looking at Locust Cycles) focuses on cicadas (commonly called locusts) having different life cycles and raises some natural questions related to these cycles. For instance, Problem 4.2 asks:

> a) How many years pass between the years when both 13-year and 17-year locusts are out at the same time? Explain how you got your answer. b) Suppose there were 12-year, 14-year, and 16 year locusts, and they all came out this year. How many years will it be before they all come out together? Explain how you got your answer.

The discussion on cicadas proceeds with a follow-up problem and is continued, with added complexity, on a later page. For example, after a brief introductory paragraph sets the stage, Extension 7.2 (a) poses the questions:

> Suppose cicadas have predators with 2-year cycles. How often would 12-year locusts face their predators? Would life be better for 13-year locusts?

While studying these materials, the PSTs authentically experience concepts they will be someday be teaching. After the collaborative groups report their findings and methods to the rest of the class, the lesson moves onto deepening the PSTs' understanding of these mathematical concepts by considering generalities and computational tools for analyzing similar problems. For instance, the following collection of ideas (along with other related notions) naturally connect to the mathematics encountered in this specific middle school curricular unit:

- Expressing the LCM of two integers and the GCD of two integers in terms of the prime factors of the given integers;
- Determining the relationship between the LCM and GCD of two integers;
- Using and justifying Euclid's algorithm to compute the GCD (and the LCM using the discovered relationship) of two arbitrary integers;
- Employing the computations of Euclid's algorithm to write the GCD of two integers as a linear combination of the two integers;
- Deriving several elementary number theoretic facts by exploiting the linear combination identity.

The development of these ideas would span several class periods and would be integrated into other topics of study throughout the course.

CLOSING THOUGHTS

My $(CM)^2$ colleagues and I are hopeful that our work will help prepare future generations of middle grade mathematics teachers to effectively utilize and implement NSF-funded curricular materials and inquiry-based instructional strategies. Teaching transcends occupation. Its best rewards are knowledge and inspiration. Teachers love to teach! For me, the words of Vachel Lindsay, a turn-of-the-century Midwestern poet, encapsulate the teaching challenge (Lindsay, 1926):

The Leaden-Eyed
Vachel Lindsay (1879–1931)

Let not young souls be smothered out before
They do quaint deeds and fully flaunt their pride.
It is the world's one crime its babes grow dull,
Its poor are ox-like, limp and leaden-eyed.
Not that they starve, but starve so dreamlessly,
Not that they sow, but that they seldom reap,
Not that they serve, but have no gods to serve,
Not that they die, but that they die like sheep.

CHAPTER 26

USING NSF-FUNDED MATHEMATICS CURRICULA TO ENHANCE TEACHER PREPARATION

Jeffrey Frykholm
University of Colorado at Boulder, Boulder, Colorado

The purpose of this chapter is to discuss the potential of using NSF-funded middle school mathematics curricula to support learning in preservice teacher education. I have used NSF-funded curricula in my work with preservice teachers throughout the last decade. Both anecdotal experiences as well as formal research projects have led me to believe that these curriculum materials can: (a) have a profound impact on beginning teachers' mathematical knowledge, competency, and confidence; (b) impact the ways in which beginning teachers think about how children learn mathematics; and (c) impact the formation of teachers' philosophies and pedagogical practices. In this chapter, I briefly explore these ideas first by providing a rationale for using NSF-funded curricular materials in the context of preservice teacher education. Second, I explain my "method" for using NSF-funded curricular materials with preservice teachers. Lastly, I share anecdotes from my experiences as a teacher educator, using the voices and

A Decade of Middle-School Mathematics Curriculum Implementation, pages 245–257

experiences of preservice teachers. In the end, I hope this chapter reveals the power and potential inherent in using NSF-funded middle grades curricular materials as a fruitful resource in the teacher preparation process.

THE RATIONALE: WHAT THE RESEARCH TELLS US

The American mathematics education community has worked hard to narrow the distance between the *vision* and the actual *practice* of reform-based teaching and learning in our schools. Yet, recent writings in our field continue to indicate how difficult it can be for many beginning teachers to fully embrace these pedagogical and curricular reforms. Typical classroom instruction in America still appears quite rigid when compared to instruction in other countries around the world (Stigler & Hiebert, 1999). Ma's (1999) captivating comparison of American and Chinese teachers reveals that many American elementary teachers do not have strong mathematical content knowledge and related pedagogical sophistication. Similarly, Ruth Heaton's (2000) courageous account of her own growth and development as a mathematics teacher points to the struggles that even *reform-minded* mathematics educators must face as they build teaching decisions around the notion that students can construct mathematical understanding.

If teachers are to embrace and foster teaching strategies consistent with the many reform-based documents prevalent today, then what knowledge of mathematics do they need? Furthermore, what kind of knowledge for teaching mathematics must they possess? As teacher educators we must also acknowledge the kinds of experiences, beliefs, and knowledge bases that beginning teachers bring to the learning-to-teach process. Thus we have two starting points for our work with preservice teachers: (a) What do we believe they need to be prepared to teach mathematics well? and (b) What do we know about the kinds of beliefs, knowledge, and experiences that they bring to the teacher education process? Thankfully, the research literature provides us with an excellent start toward understanding, and acting upon, these questions.

It has been widely recognized throughout the past 15 years that teachers must understand mathematics deeply themselves if they are to facilitate the types of discussions and handle the various questions that emerge when learners engage in authentic mathematical experiences (Ma, 1999). While the past decade has seen widespread and systematic attempts to bolster the content knowledge of teachers, we also know that many prospective teachers continue to possess weak content knowledge of mathematics (Lloyd and Frykholm, 2000; Heaton, 2000).

Teachers' conceptions about mathematics, about mathematics teaching, and about students' ability to learn mathematics strongly impact their pedagogical practices. Research has consistently illustrated ways in which teachers "translate their knowledge of mathematics and pedagogy into practice through the filter of their beliefs" (Manouchehri, 1997, p. 198). It is widely agreed upon that these beliefs are often formed and solidified well before prospective teachers enter teacher preparation programs, and continue to be bolstered through experiences in settings that do not necessarily reflect reform visions and goals. This situation presents teacher educators with a formidable challenge. We must recognize that the beliefs of preservice teachers are often *not* synonymous with the assumptions that guide many reform-based educational programs. It remains the task of teacher educators to first uncover, and then perhaps confront, the beliefs teachers may hold toward mathematics, toward teaching, and toward young learners.

Brown, Cooney, and Jones (1990) wrote, "Preservice and beginning teachers' reversion to teaching styles similar to those their own teachers used are legendary" (p. 649). This contention raises a compelling point—if teachers tend to teach the way they were taught themselves as learners, what can we expect in the classrooms of tomorrow? Reports generated during the decade of the 1990s left us with little optimism, painting a fairly dim picture of American mathematics teaching. These reports included evidence of, for example: little teacher understanding of the theory and recommendations underlying contemporary reform efforts; few instances of conceptual teaching in the elementary grades; relatively minimal use of hands-on manipulative activities; and few connections made between instructional decisions and students' conceptual understanding (Frykholm, 1996; 1998; 2000; Manouchehri, 1997). Therefore, it seems reasonable from this research to assume that typical past classroom experiences for many prospective teachers were likely devoid of the kinds of pedagogical strategies advocated in contemporary calls for reform.

We hope that research of the coming decade will depict a more positive picture as K–12 students in recent years have been exposed to more reform-minded curricular materials, are being given more opportunities to create their own understandings, and are increasingly being asked to think deeply about meaningful mathematical contexts. Will these richer experiences in K–12 settings eventually lead to better teaching as a new generation of teachers enters the classroom? That is a good question that some have already begun to pursue. For now, however, we must be aware that experiences *as learners,* in almost every context, are extremely powerful and have lasting effects.

MY METHOD: USING NSF-FUNDED CURRICULA IN PRESERVICE TEACHER EDUCATION

The previous brief summary of the research literature tells us something about what preservice teachers bring to educational experiences with respect to their knowledge, beliefs, and experiences. The next logical question remains: Given their backgrounds, *to what kinds of additional experiences and perspectives should they now be exposed?* Much recent debate about this question has arisen, particularly in recent years as new and pedagogically challenging mathematics curricula have begun to have a strong presence in the public school context. These programs often require teachers to facilitate discussions and engage in mathematical thinking and authentic problem-solving activities simultaneously with their students. For many teachers, this change in both role and purpose—a stark contrast to what they observed their own teachers doing when they were K–12 learners—is no easy shift. How might we better prepare teachers for these new tasks? The Mathematical Sciences Educational Board and National Research Council (MSEB, 1989) have contributed to this conversation, suggesting, "teachers themselves need experiences in doing mathematics—in exploring, guessing, testing, estimating, arguing, and proving... they should learn mathematics in a manner that encourages active engagement with mathematical ideas" (p. 65).

I suggest that the use of NSF-funded curricular materials in mathematics methods courses is both relevant and effective toward the vision of the MSEB. These types of curricular materials have the potential to stretch preservice teachers with respect to content, and at the same time serve as excellent launching points for discussions about pedagogy and student learning. These materials explicitly incorporate reform ideas about mathematics content and pedagogy by emphasizing real-world contexts, providing opportunities for problem solving, promoting the development and discussion of multiple solution strategies, and making connections both within and external to mathematics. In the following sections, I highlight four particularly important areas in which the use of NSF-funded middle school mathematics curricula can have notable impact: content knowledge; pedagogical content knowledge; understanding of, and belief alignment with, innovative curriculum programs; and comfort in the dynamic mathematics classroom. When viewed together, I believe these four themes constitute the beginning of a framework that may be helpful to guide the work of teacher educators as they prepare future teachers for the demands of the mathematics classroom.

Content Knowledge

In order to successfully engage students in rich mathematical thinking and reasoning, teachers must possess rich content knowledge. There have been many conversations among mathematics educators over precise definitions of content knowledge. What I suggest in my own teaching and research, however, is a broadening of the notion of "content knowledge" from simply a reservoir of knowledge and experiences to include the *capacity to acquire* the mathematical understanding and/or insight necessary in a given moment. It is my contention that using NSF-funded curricular materials in the context of preservice teacher education contributes both to the traditional "canon" of content knowledge necessary for teaching, but also to this expanded notion of the capacity *to acquire new knowledge* as contexts demand.

Pedagogical Content Knowledge

Teachers must understand the ways children think intuitively about mathematics, their native strategies and problem solving tendencies, and the importance of utilizing teaching strategies that match children's ways of knowing and learning. These forms of pedagogical content knowledge (Shulman, 1986) continue to be an instrumental element of the reform-based classroom, and must be considered a vital component of any teacher development process. One of the salient features of the NSF-funded curricular programs is the extent to which they include—indeed, are structured around—the native thinking, intuitions, and strategies of children.

It is common in the teacher manuals of these texts to see examples of student thinking, mention of common misconceptions students may develop, or examples of thought-provoking questions that are appropriate for the contexts at hand. While these teaching aids are helpful at any time, I contend that they take on new meaning for teachers who have engaged *as learners* in the mathematical content to which the pedagogical "tips" apply. It is one thing for a teacher to read about a potential misconception a student might develop. It is another for that teacher to have the very same misconception exposed in his/her own thinking. When teachers engage the NSF-funded curricular materials as learners, I believe that they acquire a new level of pedagogical content knowledge that is likely to be achieved only superficially without experiencing the contexts as learners.

Understanding of, and Belief Alignment with, Innovative Curriculum Programs

Innovative curriculum programs are often based on philosophies of mathematics education that differ significantly from those that have guided curricula in the past. Therefore, effectively teaching NSF-funded curricula requires that teachers understand the principles upon which the programs are based, as well as implicit assumptions these programs make about the changing roles of teachers and students. Teachers need opportunities to examine their beliefs about the nature of mathematics teaching and learning as they come to understand the assumptions that underlie these types of mathematics curricula. My experiences with these materials over the past decade suggests that the most legitimate and compelling way to confront preservice teachers' beliefs about the nature of mathematics teaching and learning is to engage them in authentic mathematical activity. In so doing, preservice teachers experience for themselves the power of multiple solution strategies, the relevance of multiple representations, and the value of discussing mathematics.

Comfort in the Dynamic Mathematics Classroom

The fourth principle in this framework includes teachers' ability to manage effectively the dynamic classrooms demanded by reform-based curriculum programs. With the curriculum programs of today, teachers can find themselves in situations where the well-established roles of both teacher and student are challenged. Problem contexts that elicit multiple solution strategies require teachers not only to *value* competing viewpoints, but to connect and/or address these various mathematical ways of thinking in the natural flow of classroom discourse. These kinds of situations necessarily give rise to moments of uncertainty in the classroom that, until recently, have mostly been suppressed by teachers as quickly as possible.

Borrowing from the discipline of literacy, Susan Villaume (2000) has helped us begin to think about the importance of this issue, noting in particular the "terrible freedom that some teachers experience as they realize they are being asked to stop teaching programs, and start teaching based on what children are thinking and doing" (p. 19). Similarly, Stephen Brown reminds us of the value of uncertainty as a pedagogical principle when he discusses "reconstructed visions of teaching that... appreciate what it means to be confused, what virtues may follow from it, and what pedagogy may make use of it" (Brown, 1993, p. 108). These scholars have articulated the importance of allowing beginning teachers to experience the kinds uncertainty that are common to the reform-based classroom early in their professional education. This experience includes the discomfort students might have as they encounter open-ended contexts and share ideas with peers, as well as the uncertainty that teachers inevitably face as they guide students in a dynamic setting.

The use of NSF-funded curricular materials in the context of preservice teacher education provides useful exposure to the complexities that await beginning teachers. Moreover, properly modeled, preservice teachers can experience (perhaps for the first time) how an implementation of these kinds of materials might be accomplished. Recall the earlier noted finding that teachers tend to replicate the very same strategies that they experienced as learners. Subsequently, it seems both obvious and critical that preservice teachers have positive experiences with these materials prior to being asked to implement them as teachers.

Some Nuts and Bolts About Using NSF-Funded Mathematics Curricula

The primary assertion I make in this chapter is that the use of NSF-funded middle school mathematics curricula can be a central component in the process of preparing teachers for the demands of the mathematics classroom. In the following sections, I briefly share a few ways in which these materials can be implemented effectively in the context of the mathematics methods classroom, and in so doing set the stage for the anecdotes of practice that follow. In particular, I discuss the attributes of problems from the curricula that are well-suited for the methods classroom, and how I use them as focal points of discussion with my students.

Selecting problems. Hundreds of problem contexts exist to choose from in a program like *Mathematics in Context* (*MiC*) (National Center for Research in Mathematical Sciences Education & Freudenthal Institute 1997,1998; Wisconsin Center for Education Research & Freudenthal Institute, 2006), or any other of the four NSF-funded middle grades mathematics curricula. Which ones will lead to fruitful engagement and discussion in the methods class? Three essential qualities guide my choice of problems to use with preservice teachers.

First, the mathematical content in the problem context itself must be challenging for the preservice teachers. One might think that this would be the most difficult challenge—selecting a problem designed for 5th–8th grade students that will also be intellectually stimulating for prospective teachers. In reality, however, this is not so difficult. Many of my students have never been asked to think in ways demanded by these curricula; the mathematics in these programs is often challenging for them. Finding a problem that is intellectually stimulating for 25–30 students means that the problem will likely be contextually-based, allowing for the preservice teachers to enter and exit the problem in various places. (This choice is strikingly similar to the choice of a middle grades teacher who seeks to engage all students in mathematical inquiry.)

Second, and somewhat related, I always look for problems that can be solved with more than one strategy. This characteristic is essential for two reasons. First, as a teacher educator, I find that the amount of time I have in a methods course is always too short; I feel that I must take advantage of every opportunity to model best practices. If we, as mathematics teacher educators, believe that multiple-strategy use is important for the public school classroom, then we must find ways for our students to experience the benefits of multiple strategy use personally so that they will seek such contexts in their own teaching. Second, when multiple solution strategies are shared, the possibility for rich discussions about mathematics emerges as the preservice teachers compare, agree, disagree, discuss, and make connections among mathematical concepts. As noted previously in the framework for this article, one of the great challenges facing teacher educators is to help preservice teachers develop their content knowledge of mathematics. As I seek to address this deficiency, I try to take advantage of opportunities to engage my students in discussions of mathematics. I have found that these discussions are greatly enhanced by problem contexts that elicit more than one fruitful strategy.

Third, I often try to select problems for which I also have access to samples of middle school students' responses to the problem. I have two strategies for collecting student work. Many of the teacher materials for the curricula include samples of student thinking that may, for example, highlight common misconceptions about, or common strategies used for, a particular problem context. I have often collected samples of student work myself—in various research contexts, but also informally with children that I know out of school contexts.

It is important to use samples of middle school students' work in connection with these problem contexts for several reasons. First, its use directly addresses skeptics who regularly say something like: "Sure, *we* came up with these solutions as college students, but middle school kids would never do it like this." In my experience, the middle school students' strategies are often more creative and elegant than my preservice teachers' strategies! Second, it is often an eye-opening experience for the preservice teachers to recognize just how powerful the thinking of children can be. Finally, samples of student work provide ample opportunity to extend the conversation from the mathematics itself to discussions about pedagogy.

Fostering discussion about teaching and learning. Once the problem context has been selected, a whole host of issues emerge related to effectively implementing the problem in the methods classroom. As noted earlier, I believe we must take every opportunity available to model best practices for K–12 mathematics education. Toward that end, I usually follow a similar process every time I use these kinds of problems. After an ample (but not extensive) introduction to the problem context, I begin by letting students work together at their tables on the problem. Sometimes this takes 10 minutes; often it will take longer. It is imperative that the students have enough time to fully engage in the problem. Next, I allow several groups to present their thinking and strategies in front of the rest of the class. Third, I allow

Figure 26.1 What is the cost? What is the cost of a soda? Of a T-shirt?

time for general discussion about the students' solution strategies, including specific attention to assumptions the students made, differences and similarities among strategies, and ways in which the strategies may be connected. Finally, when possible, I present the solution strategies of middle school students, which often provides a space for further conversation as we highlight the ways in which young students, without the "mathematical socialization" that often clouds the thinking of adults, are free to think deeply and creatively about mathematics. Throughout, I look for opportunities to highlight various pedagogical strategies that might be relevant for any given problem context.

An example. Figure 26.1 contains a problem that I borrowed from the *MiC* unit *Comparing Quantities* (Kindt, Abels, Meyer, & Pligge, 1998). This problem allows students to develop, among other things, the conceptual foundation for solving systems of equations. While most of my students apply traditional algebraic processes toward the solution (solving two equations with two unknowns using the elimination or substitution methods), they are regularly shocked when they see the elegance of the solution strategies presented by middle school students (see Meyer, 1999, for a discussion of student strategies to a similar problem). It is always quite easy, therefore, to use this problem to stimulate conversation not only about mathematics itself, but how we might effectively teach it to middle school students.

ANECDOTES OF EXPERIENCE

Perhaps more convincing than research findings and theories presented above are the simple words of preservice teachers who have experienced these NSF-funded curricular materials in the context of their teacher preparation program. As I share the voices and anecdotes of my students over the years, I recommend that the reader remain cognizant of the theoretical framework of this chapter that highlights the ways in which these anecdotes illustrate how NSF-funded curricula can help teacher educators address the content knowledge of beginning teachers, their belief structures, their understandings of pedagogy, and their emerging images of themselves as math teachers.

Starting Points: Mathematical Autobiographies

The first assignment that I give my students is to write a short autobiography about their journeys as learners of mathematics. These accounts commonly include stories of indifference as many students have very few recollections of their early math experiences. As one student noted, "I know this is really bad to say, but I just can't remember much of anything about math in elementary school. I know that we did it, but for the life of me I can't tell you anything about it" (Julie, 2000, journal entry).

More common than the many stories of ambivalence, however, are narratives that contain powerful statements of fear, uncertainty, and failure. As one woman noted, "My experiences in math were so awful that even writing this paper has made my stomach tighten in a knot, and my hands shaky. I haven't forced myself to think about math for years, which is one of the reasons I am so terrified about taking this course." (Alison, 2000, journal entry).

The issues that surface in the writing of the mathematical autobiographies represent the beliefs many of my students bring to the course prior to engagement with the curriculum materials, as well as their philosophical stances toward reform ideals that are based largely on their experiences as learners in the K–12 math classroom. Most of these views are not positive.

Extending Mathematical Content Knowledge

One powerful characteristic of using NSF-funded curricular materials with preservice teachers is that they often lead these students to recognize the holes in their own knowledge, particularly as it relates to conceptual understanding. As Allen (2001, interview) stated, "I thought this math course was going to be an easy, simple refresher. Never would I have suspected how little I understood the basics of mathematics." Another student shared similar sentiments:

> The curriculum activities we have used in this class have completely challenged my own knowledge and understanding of mathematics . . . not because the mathematical material was new, but because I had never tackled it in these kinds of ways before, which makes me realize how little I actually know about math! (Julie, 2000, interview)

It appears that engagement with the curriculum materials challenges not only *what* preservice teachers may know about mathematics, but also *how* they know mathematics. "The most challenging part of the class," said one, "has been reexamining the ways that I learned to do math, attempting to put aside what I know in order to try these new methods. It is incredibly difficult to set aside the methods that have been drilled into my head (Sarah, 1999, interview).

This tension often arises for students who find themselves in unsettling territory—caught between the habits of mind that have been their mathematical tendencies for many years, and a new (for them) way of thinking conceptually about mathematics that is still in its infancy. This tension between reconciling old habits with emerging conceptions and understandings becomes a regular topic of conversation. As one student contributed:

> I find myself in new and rather unsettling territory. I have learned that I am simply a product of successfully memorizing the algorithms and not truly understanding mathematics... I keep wondering if I can successfully teach mathematical concepts that I have been struggling with myself. I am trying to unlearn much of what I have already learned and start over. (Sarah, 2002, interview)

Engaging Preservice Teachers' Conceptions About Mathematics Learning

My students often comment on the parallel between their own learning and the learning of children. As Erika (2000, interview) shared, "I definitely have a better understanding now of how children obtain not just knowledge of mathematical ideas, but construct their own understandings of mathematical concepts and actually think mathematically." Other students elaborated on the connections between the materials and the mathematical thinking and varied solution strategies of children. Kevin (2003, interview) reported,

> Children, and adults, learn mathematics in many different ways. At first, I was surprised when other people in the class had entirely different approaches to the problems we were asked to solve. I always thought my way to solve the problem was the "right" way. However, that theory went out the window after the first hour in your class. It was obvious that everybody in our class used different approaches to solve math problems. This is true for children as well. There is no right way or wrong way to solve problems. As teachers, we need to understand ALL these approaches!

Other students extended this idea that curriculum materials can encourage multiple solution strategies by adding that, through rich contexts, children can *teach themselves* meaningful mathematics. This recognition that children can "learn on their own" through rich problem contexts provides an excellent opportunity to engage students in discussions about children's *opportunities to learn*, as well as the notion that the *kind* of mathematical understanding students will likely develop is a product of the classroom experiences they receive.

Confronting Conceptions About Mathematics Teaching

At some point, most beginning teachers must confront their own discomfort and uncertainties caused by teaching mathematics in a way that diverges from their experiences as learners. For some students, the connection between learning and teaching becomes intuitively obvious. As Dana (2002, interview) questioned, "The idea of teaching math conceptually now makes so much more sense.... Why *wouldn't* you teach in a way that actually utilizes what children know and, as we discussed in class, matches 'children's natural, intuitive approaches to mathematics?" A number of students made similar comments, including those who note that good mathematics teaching may mean that the teacher should actually say and do *less*.

> My engagement with the curricular activities in this class, as well as the reinforcement of them in my practicum classroom experience, have significantly altered my thoughts about how best to teach mathematics to children. I guess the main thing that I have extrapolated from this class and my practicum is not to explicitly teach kids anything that they can figure out themselves through *doing*. (Jeff, 2001, interview)

When juxtaposed with the strong statements of anxiety that many students express at the beginning of the course, I am often startled at the confidence with which students speak about their emerging conceptions of mathematics teaching by the end of the course. They discuss the importance of wrestling themselves with mathematical ideas, as well as the importance of having those struggles as contributing members of the mathematical community developed during the course. As one student shared,

> The curriculum materials had a tremendous impact on my thoughts about teaching. Before this class I was only aware of the traditional rote, mechanical approach to math. This class has given me many tools to use so kids can better understand mathematical concepts and also have an easier time doing so. The traditional method of teaching math has pretty much gone right out the window for me, and been replaced by a much more child-centered approach to mathematics. (Kevin, 1999, interview)

SOME FINAL THOUGHTS

I have no doubt that giving preservice teachers the opportunity to engage *as learners* in NSF-funded middle school mathematics curricular materials can stimulate growth in their own content knowledge of mathematics, as well as their understanding of teaching and learning processes. Of course, curricular materials alone are not the only answer. The use of NSF-funded curricula in teacher preparation experiences, however, seems to hold promise for more than just developing mathematical understanding. What I have seen emerge numerous times is a rich context for learning simultaneously about mathematics, about the thinking of students in mathematics classrooms, and about pedagogy. In the end, the dynamism of the conversations that emerge in these contexts may in fact be the most powerful agent in the process. Certainly, as provocative as they may be, curricular materials alone are just print on paper. They only come alive as they become a focal point of debate, conversation, and reflection. With that consideration, I close this chapter with one final quote from a student that captures the essence of my conviction that these materials are not only worthy of use in the context of teacher preparation, but of great significance as well.

> I came into this class with reservations not only about my ability to effectively teach math but also about my own level of math proficiency. I was fully expecting to hate [this class]. But, I have learned a great deal.... I think it is great that we are using materials that actual kids would be using in good math classrooms, because most of us never had the opportunity as children to learn math in these ways. Had I not taken this class, I probably would go into the classroom and teach the way that I was taught: right out of the book, algorithmically. When children would ask me "Why?", I probably would have said what all of my teachers used to say to me: "Because that's just how you do it. Math has certain rules you just have to learn." How many children are left behind by that approach? I used to feel very intimidated that question. But, because of this class, I now feel much more ready to tackle math in my classroom. (Julie, 2000, interview)

Perhaps these words resonate with you as they have with me, and they will encourage us to continue to pursue ways in which innovative *middle school* materials can be used as a key component for reforming the *university*-based preparation process.

CHAPTER 27

NSF-FUNDED MATHEMATICS CURRICULA IN THE FIELD

A Preservice Teacher's Opportunity to See Theory in Practice

Ann Bledsoe
Columbia College, Columbia, Missouri

One of the major issues concerning mathematics teacher educators today is preservice teachers' past experiences and beliefs. Cooney (1994) suggests that preservice teachers' orientation towards mathematics teaching and learning "appears to be rooted in experiences that occur long before teachers begin their formal mathematics education courses" (p. 12). He suggests that this orientation is often one in which teachers treat "mathematics as a cut-and-dried subject," (p. 12) which is not what educators who subscribe to a reformed vision of mathematics teaching and learning hope to portray in their mathematics methods courses.

A second issue of concern for mathematics teacher educators is the alignment of theory being introduced and encouraged in mathematics methods courses and many of the teaching practices seen by the preservice teachers during their field experiences. Vacc and Bright (1994) suggest that if "nov-

A Decade of Middle-School Mathematics Curriculum Implementation, pages 259–266

ice teachers get too little practice or if they receive inconsistent feedback from their university teacher educators and classroom supervisors, their expertise may not be adequately developed" (p. 117).

In this chapter, I describe how reformed-based middle school mathematic curricula can be used as tool to help preservice teachers see the theory proposed in their mathematics methods classes come to life in a middle school mathematics classroom. In doing so, I present a preservice teacher's viewpoint of how the theory she was learning in my middle grades mathematics methods course became reality in her field experience.

PRESERVICE TEACHERS' PAST EXPERIENCES AND BELIEFS

The issue of preservice teachers' past experiences and beliefs is indeed a big concern for teacher educators. Borko and Putnam (1996) outline the issue.

> New teachers are likely to bring to their initial teaching experiences a host of assumptions about the nature of learners and learning, assumptions that shape the instructional skills and routines that they learn. In many cases, these beliefs about how students learn and the teacher's role in facilitating learning—beliefs acquired over years of experience as students in traditional educational settings—are incompatible with the views of learning underlying the instructional approaches advocated by teacher education programs. (p. 699)

Further, Borko and Putnam argue that these pre-existing beliefs "exert a powerful influence on what they learn about teaching" (p. 701). They suggest that programs that want to challenge these beliefs must help preservice teachers "make their implicit beliefs explicit" and "create opportunities for them to confront the potential inadequacy of those beliefs" (p. 701).

Thus it falls to mathematics educators, who adhere to reform mathematics ideals, to provide opportunities in their courses for preservice teachers to overcome obstacles created by their past experiences and beliefs. However, even if teacher educators make headway against these past experiences and beliefs in their courses, there is no assurance that preservice teachers will be able to maintain this new view when faced with the reality found in many classrooms.

The Field Experience

The reality of mathematics teaching and learning in many middle grades classrooms is another area of concern for teacher educators. This concern draws from the frequent misalignment of the theory proposed in their mathematics methods courses with the often traditional teaching practices encountered by preservice teachers during their field experiences. Zeichner (1993) states that "[s]tudent teachers throughout the world are frequently placed in classrooms where the teaching they are exposed to often contradicts what they are taught in the colleges" (pp. 10–11). Panasuk, Cutler, and Sullivan (1998) also suggest that "student teachers are often assigned to classrooms spontaneously" and "some of the lessons they observed are of dubious quality" (pp. 17–18). They feel that "[t]his practice undermines the goals of the teacher preparation program and interferes with prospective teachers' development because what they observe is at odds with their knowledge about pedagogy" (p. 18).

Now the issues become intertwined. If teacher educators are able to overcome the traditional view of mathematics held by many preservice teachers through measures taken in their methods courses, this may be undone by the experiences had by the preservice teachers in the field. Frykholm (1996) suggests that this perceived undoing would arise because "student teachers conform to existing conventions and practices even though they are inconsistent with personal beliefs and values (pp. 676)." That is, if the view of mathematics teaching and learning in the field experience is traditional, then although the preservice teachers' beliefs may have started to change during their methods course, they will not stand up in the reality of the classroom.

How can a mathematics teacher educator attempt to overcome the past beliefs and experiences held by preservice teachers, especially when these beliefs and experiences are in conflict with the beliefs about mathematics teaching and learning that many educators hope to portray in their courses. What experiences can teacher educators provide to convince preservice teachers that different beliefs are sound not only in theory but also in practice? That is, if teacher educators are able to convey the reformed view of mathematics teaching and learning during courses, will preservice teachers be able to see and apply this view in an actual mathematics classroom?

The following two stories provide a view of how reform-based mathematics curricula can play a role in helping mathematics teacher educators portray and promote the views of the reform movement in their courses and in helping preservice teachers see such views enacted in actual mathematics classrooms.

A TEACHER EDUCATOR'S EXPERIENCE

The preservice teachers in my middle-school mathematics methods course were required to read a good portion of the *Principles and Standards for School Mathematics* (NCTM, 2000) and class discussions were held about what they had read. In addition, they were also asked to read and discuss material regarding pedagogical issues, such as various learning theories, including the learning theories reflected in the *Principles and Standards for School Mathematics.* In this way, I felt that the preservice teachers had been given the opportunity to get a basic understanding of the ideas about teaching and learning mathematics proposed by the reform movement.

I wanted my students to see these ideas in practice as both learners and future teachers of mathematics. As Jeff Frykholm discusses in the previous chapter, the use of a standards-based middle school mathematics curriculum can be one way to present these ideas to preservice teachers in a methods course. I routinely used activities from *Connected Mathematics* (Lappan, Fey, Fitzgerald, Friel, & Phillips, 1998, 2002, 2004, 2006), *Mathematics in Context* (National Center for Research in Mathematical Sciences Education & Freudenthal Institute, 1998; Wisconsin Center for Education Research & Freudenthal Institute, 2006), MATH*Thematics* (Billstein, & Williamson, 1999a), and *MathScape* (Education Development Center, Inc., 1998, 2005) to engage my students in thinking about mathematics pedagogy and to learn some mathematics in the process. I was pleased to see that they saw the value of these activities and they often informed me of their intent to use them, or similar activities, in their own classrooms in the future.

Each semester of my course, students were required to teach a short lesson which addressed a particular content standard. When planning these lessons, the students were given access to the four NSF-funded curricula and many chose to use activities from these materials. Their choices of lessons from these materials, as well as their reflections on these choices after completing the assignment, led me to believe that the students agreed with the principles of the reform movement and saw the activities in these materials as a way to help them teach in a way that was aligned with these principles.

I felt that these experiences with the four curricula provided opportunities for my students to construct both a theoretical knowledge and a practical understanding of the ideals set forth in the *Principles and Standards for School Mathematics* (NCTM, 2000). While in my class, many of the students embraced this view of mathematics teaching and learning as their own. As their instructor, I knew that one test of this shift to a new paradigm would be the instructional choices that they would make in their field experience. Fortunately, these field experiences were in a school district that had recently adopted *Connected Mathematics Project* (CMP). As such, they were able to walk into a classroom full of students using these materials and be given the opportunity to teach a lesson from these materials. Because of this, many of my students were able to validate and solidify their new found beliefs about teaching and learning mathematics. The story of one such preservice teacher illustrates this point.

A PRESERVICE TEACHER'S REALIZATION OF THEORY IN PRACTICE

Erin was a student in my methods course during the fall of 2000. She came into my mathematics methods course with a background in mathematics and ideas about teaching that primarily came from traditional instruction from kindergarten through high school. Early on in the semester she informed me that my course was introducing her to a whole new world of mathematics, yet, before she could change as a teacher of mathematics, she had to first change as a learner of mathematics.

In my course, students were asked to share their mathematical autobiographies. Erin's story was typical of those of her classmates. From an early age, Erin was one of the students who could follow the "prescribed algorithms of mathematics." She professed that she would make up songs and rhymes to remember steps to mathematical procedures. Unfortunately, she found, as she progressed though mathematics in junior high and high school, that if she did not remember these rhythmic guides, she felt helpless because she did not know the "why" behind the steps. She struggled with higher-order thinking tasks if she did not immediately know the procedures to apply.

Luckily, Erin was fortunate enough to have a few teachers in high school who promoted the idea that doing mathematics is more than just memorizing a list of rules. They challenged her to analyze her thinking, write out her steps, and consider logic in her computations. She was encouraged to think in terms of concrete representations and to consider the reasonableness of her results. These few experiences left Erin primed to want to teach mathematics to her future middle school mathematics students in a similar fashion, but unsure of how to do so. One of her reflection papers from the middle of the semester expressed her beliefs:

> As I look back now, I can easily pinpoint my high school experiences as priceless because of the bridge they began to build in my mind between steps and meaning. When I arrived in this middle school math methods course, I did believe that math was more than memorizing steps. I also believed that the teacher's role in math was not to spoon feed the students algorithms and rules. Even if presented creatively, they were still just steps that locked a student in when pressed to manipulate their information with situations slightly different from the norm. As a student, I had seen many lights come on in my mind as meaning unfolded before me related to steps that I had long since forgotten in my past. I can finally explain why, when dividing fractions, you flip the second fraction and multiply. It was not that these previously memorized steps become void, but instead, they became authentic, personalized, and fluid, to be used for my own mathematical inquiries. Despite all of this, as a preservice teacher, I still am not convinced that these ideals, as seen and supported in the *PSSM*, will really work in a classroom of authentic students.

As part the teacher education program at the University of Missouri-Columbia, my students were given the opportunity to observe and assist professional teachers and their students in area schools. The district in which they were placed had recently adopted reform-based mathematics curricula for all levels K–12. Erin had the opportunity to visit mathematics classrooms at different grade levels through various methods courses besides my own. She reflected on these field experiences in another paper:

> As I went from experience to experience, I was amazed from the beginning at how math had changed since I was in elementary school and junior high. No longer did I see the thick, hardback books filled with pages and pages of practice problems. Instead, students were working with thinner books that had explorations and investigations with numbers. There were brightly colored illustrations and countless demonstrations with models and manipulatives that showed and explained the concepts. Students were encouraged to use pattern blocks, base-ten blocks, pictures, diagrams, and words to explain and justify their findings. Unlike my memories of grade school where I would be given 50 problems to complete individually at my desk, these students were given only two or three problems so solve in as many ways as they could either on their own or in small groups. Before a hand would go up to ask the teacher a question, students used each other as resources for information and critically questioned the logic that was offered. I could barely pull myself away from these math classes and felt excited and motivated to learn about how to create such a learning environment and facilitate this kind of learning as a future teacher. My skepticism still lingers, but more than ever I am beginning to see for myself what true mathematics could and should look like.

According to Erin, one of her most powerful experiences with standards-based curriculum came when she was given the opportunity to teach an activity in a sixth-grade mathematics class. The students were using *Prime Time* from *CMP*. The objective of the activity was for the students to identify factor pairs for whole numbers. In her lesson reflection, she wrote about her first reaction after reading the activity when planning her lesson.

> When I first approached the content, my background as a student initially kicked in. It would have seemed very time efficient to simply tell the students about pairing two numbers together to get a product and labeling them factor pairs. We could have then gone to memorizing the facts and drilling on hundreds of examples. However, if asked to tell why they were doing these steps, I am confident that few, if any, students would be able to pinpoint the "why" behind what they were "learning." In reality, this "time saving" method would have only kept them from moving on to more advanced thinking in the future.

Erin opted for the approach suggested in the curriculum in which the students were told that 12 desks needed to be arranged in a rectangle for the most efficient use of space for an upcoming exhibit. The students were given grid paper and allowed to use square tile manipulatives to find all of the different rectangular arrangements using 12 squares to represent the desks. With this established, Erin let the students work individually at first and later with a partner. While they did so, she walked around and interacted with the students. She encouraged them to write down their ideas and then test them to see if they worked. She would tap students on the shoulder, compliment them on their thinking, and ask them if they would be willing to share their thoughts when the whole class discussion began. When the whole class came back together, the first thing she established was that she did not want just the answer. Instead, she told the students to begin by sharing their strategies and ideas. Students came to the overhead to draw their models and others simply explained from their seats. After several minutes, the "answer" was clear to the students, yet they were still quite interested in the ways their fellow classmates had found this answer.

Erin continued the investigation by suggesting a new amount of desks—16. Students again went to work to find the possible rectangular arrangements. As the students worked in their groups on this extension, they said things to each other such as, "If you just multiply the dimensions of the rectangles together, you get the number of desks possible." Students then pointed out that the dimensions of the rectangles were the factors of the total number of desks. When the class regrouped, and shared their results to this problem, Erin introduced the term "factor pairs." As she did, students reacted by saying, "this is so easy because all you had to do was think in terms of desk arrangements."

After completing the lesson, Erin shared her feelings about the lesson in a reflection paper.

> I feel that I clearly saw first hand how the power of the ideals in the *PSSM* impact students as learners and how they have impacted me as a preservice teacher through the use of the standards-based curriculum. I experienced how these students were able to explain and support their reasoning and their answers in multiple ways, along with critiquing the strategies offered by their peers. The learning that took place in this math class far surpassed that which would have [happened] through pages of calculations. These students grappled with their own learning, gave and received authentic feedback from their peers, and claimed ownership of key mathematical concepts.

In addition, Erin reflected on how she felt the use of reform-based curriculum enabled her to teach in the way that would reflect her recently formed view of teaching and learning mathematics.

> As a teacher, I was able to create a learning environment sensitive to student needs and questions, and provide personalized feedback due to the communication of their thought. Rather than being confined year after year to the same steps of administering math, the structure of the standards-based curriculum enables me to become a dynamic facilitator personalized to the students that I have at that time. Teaching can be maintained as an ever-changing position that requires thinking and processing to challenge individualized learners.

CONCLUSION

I strongly believe that the use of reform-based middle-school mathematics curricula supports the enactment of the ideals of the *Standards* in the classroom by preservice teachers. Support for these ideals begins in courses at the university level that promote the type of mathematics learning and instruction envisioned by the *Standards* and is given authentication with successful field experiences in which they see this type of learning and teaching in action. Ball and Cohen (1996) agree that "teachers could be engaged with curriculum materials in ways that generate learning if the materials were integrated into a program of professional development aimed at improving their capacity to teach" (p. 8). By modeling for preservice teachers on the college campus use of reform-based curricula mathematics teacher educators begin to revise many preservice teachers' previous experiences with traditional mathematics education. Then, as they engage in field experiences as described in this article, further links are established in the preservice teacher's schema of what reform mathematics is and they are empowered to teach mathematics in ways that support these new found ideals.

PART IV

WHAT HAVE WE LEARNED AND WHERE DO WE GO FROM HERE?

CHAPTER 28

LESSONS LEARNED FROM THE SHOW-ME PROJECT

Barbara Reys
University of Missouri, Columbia, Missouri

The chapters in this volume tell the story, played out over a 10-year period, of an effort to change and improve middle school mathematics programs through standards-based curriculum investigation and implementation. The Show-Me Center, along with three other national Centers (ARC, COMPASS, and K–12 Mathematics Curriculum Center), extended the work of 13 curriculum development projects (Hirsch, 2007) by providing support to teachers, schools, districts, and states to review and implement new models of curriculum materials based on the National Council of Teachers of Mathematics (NCTM) *Curriculum and Evaluation Standards for School Mathematics* (1989). The curriculum development efforts, coupled with extended support for curriculum investigation and implementation, represent a unique and long-term commitment to school mathematics improvement by the National Science Foundation.

The curriculum development work could not have been possible without a clear vision for school mathematics. The vision articulated by NCTM was bold and represented a major change for many teachers and schools. Yet many were eager to respond to the vision, acknowledging that school mathematics programs needed to change to reach all students and to pre-

A Decade of Middle-School Mathematics Curriculum Implementation, pages 269–277

pare them for a changing world. Many also realized that new and different mathematics curriculum materials were needed to bridge the gap between vision and current practice.

Looking back over the past 20 years of vision setting, curriculum development, and curriculum implementation reveals progress and obstacles. This volume is an effort to document some of the important lessons learned. This chapter summarizes and reflects on some of the key messages told by the authors of the early chapters. They are further discussed in the chapter that follows by Meyer and Langrall.

ADOPTION AND IMPLEMENTATION

The authors of the chapters in Part II of this volume told stories, often personal ones, that convey lessons learned about changing the culture of mathematics instruction within a school district with textbook adoption as the impetus. Each story was told by a primary player within a particular setting, lending authenticity to the messages. The chapter by Swafford and Langrall, (Chapter 12) provided a "reality check" for this volume, reminding the reader that the true potential of standards-based textbooks is not automatically realized upon adoption and implementation. Some common messages drawn from the set of chapters in Part II are summarized here and organized around the two themes of the Part.

Review and Selection of Standards-Based Mathematics Textbooks

In many school districts, the review and selection of textbooks is short-changed and often involves a small number of people. In previous eras when most textbooks were alike, perhaps this was appropriate. However, in the current era many different kinds of textbooks are available and they differ dramatically in their philosophical and pedagogical orientation (Reys, Reys, & Chavez, 2004). Therefore, the review and selection of textbooks at a district or state level is more complex and has greater consequences for users of the textbooks (students and teachers). In today's curriculum environment, the review and selection of mathematics textbooks must be elevated in importance. Careful planning at this stage can provide the impetus for significant and widespread change in classroom practices.

The stories in this section underscore the value of curriculum investigation as a vehicle for professional growth. That is, the review of mathematics curriculum materials, which generally occurs every 6–8 years in school districts, can provide an opportunity for teacher professional growth and renewal. However, without careful planning the potential benefits of this process will be unrealized. Two stories (Chapter 7 by Hoover and Chapter 4 by Tuckey) described the role and influence of a coordinated NSF-funded project on district adoption processes. As director of a statewide NSF project, Hoover emphasized the power of collaboration and community growth that resulted from representatives from many school districts working together for a common goal. Likewise, Tuckey acknowledged the support provided by a project that brought together many school districts, each of which eventually made its own decision about curriculum materials. These stories are as much about the value of NSF-supported teacher enhancement initiatives as they are about selecting curriculum materials. Collaboration among schools adds layers and levels of discussion among teachers and administrators that are unique and powerful. Although powerful, collaboration across school districts in the selection of curriculum materials is a rarity. Rectanus (Chapter 5) and Nielsen (Chapter 6) described carefully planned local processes that provided the necessary support for major curriculum innovation in their districts.

Collectively, the authors of these four chapters conveyed "lessons learned" about the review and selection of mathematics textbooks including:

- Textbook review (what some authors call curriculum investigation) is a powerful vehicle for professional growth and school improvement efforts.
- Establishing a shared vision/philosophy of mathematics teaching and learning is essential at the early stages of curriculum review.
- Engaging as many teachers within a school or system as possible provides an opportunity for "ownership" and collaboration.
- Involvement of administrators and parents is essential.
- Selection of new curriculum materials must include consideration of the potential professional development needs of teachers.

Implementation of Standards-Based Mathematics Textbooks

Several chapters in Part II conveyed how particular school districts went about implementing new standards-based mathematics textbooks and what lessons were learned along the way. Several "critical elements" for successful implementation were highlighted, including: teacher collaboration; importance of key leaders; parental involvement and support; and a coherent K–12 curriculum plan.

Teacher Collaboration. Sharp (Chapter 8) emphasized the need to get teachers from a variety of schools within a district to talk with each other. As he indicated, "the teachers did not know each other very well, but more importantly, they did not have anything they needed to talk to each other about." However, the new mathematics curriculum materials provided a focus for discussions and exchanges of ideas. Opportunities to learn the new curricula became important to teachers. As they began to use the materials, their discussions with teacher colleagues became richer and more practical. These experiences fueled small changes in practice and built confidence. They also began to view their own classroom experiences from a broader perspective as they heard similar stories, challenges, and strategies voiced by colleagues.

Leadership. Strong leadership greatly enhances teachers' success implementing standards-based textbooks. Early and continuous support from administrators is particularly critical for teachers who otherwise may be less willing to try new materials and/or instructional ideas. A leadership team (administrators and teacher leaders) helps provide an infrastructure that encourages teachers to initiate and implement change and builds the "grit" to continue if and when roadblocks are encountered. Hoover (Chapter 7), Rectanus (Chapter 5) and Nielsen (Chapter 6) indicated that "teachers on special assignment" (full-time mentors) can play key roles in supporting colleagues and communicating with administrators and parents.

Parental Involvement and Support. Virtually every author warned that parental concerns should be anticipated and addressed. A small vocal group of parents can serve as a strong catalyst or formidable obstacle to reform at the classroom, school, and district level. It is natural and healthy for parents to question a change in emphasis. Parents want to know the rationale for and consequences of proposed change and how their children will be affected. A proactive plan enlisting the support of parents and helping them understand the nature of the reform is essential.

K–12 Curriculum Articulation. Den Otter (Chapter 9) made a strong case for curriculum articulation. She wrote, "If I had it to do over again, I would beg my colleagues in the elementary schools to move more quickly to a curriculum that would support what we were doing in the middle schools." She argued that focusing on conceptual development is one of the hallmarks of the NSF-funded mathematics curricula and that conceptual development needs to be a regular and sustained focus across the grades. Efforts to provide a coherent curricular vision from elementary to middle school and from middle school to high school are rare. Yet, inherent barriers exist, particularly at the middle grades, when there is not a coherent vision. As middle school teachers implement standards-based mathematics curricula, they frequently raise important questions about the transition of students to high school mathematics, such as:

- Will our middle school students be prepared for studying high school mathematics?

- Do our high school mathematics teachers value the skills and understanding emphasized in the middle school mathematics standards-based curricula?
- If our middle school students attend a high school that emphasizes a traditional skill-based curriculum, will they be successful?

These issues need to be addressed to assure all constituents (students, teachers, parents, administrators, and school boards) that the middle school mathematics program is consistent with the secondary program. Thus high school teachers must be involved in the conversation, and high school mathematics programs must be aligned with standards-based middle school programs.

CURRICULUM-FOCUSED PROFESSIONAL DEVELOPMENT

The authors of the chapters of Part III agreed that professional development embedded in curriculum review and implementation must include attention to mathematical knowledge, understanding of inquiry based instruction, and assessment strategies that align with inquiry-based curriculum and instruction. They also emphasized that professional development sessions should involve significant opportunities for teachers to be students, using the materials to come to understand the development of mathematics and the role and nature of inquiry-based instructional strategies. The role of the professional development leader is to model good teaching practices, thereby simulating the classroom environment. Another common message among the authors was a call to use examples of student work as contexts for teacher learning. Successful curriculum-based professional development includes opportunities to establish a local community of learners where mentoring and networking provide support for ongoing curriculum implementation. Finally, the curriculum developers emphasized the need to help teachers understand important design features of the curricula.

In their chapter, Meyer and Arbaugh (Chapter 20) shared a strategy for engaging school administrators, helping them understand the need for change and the role of standards-based curriculum as a catalyst for change. They described an "Administrator Workshop" that emphasizes some of the same messages illustrated in the other chapters in this section including engaging administrators in a mathematical task to illustrate mathematical thinking and reasoning as core learning goals. Through these examples, administrators (like teachers) come to view mathematics and teaching in a different light and thereby can better understand the changes occurring in their classrooms

Personal Stories of Professional Growth

The four chapters written by classroom teachers told stories of personal growth as teachers of mathematics. Aside from the potential of NSF curriculum materials to stimulate student learning, these stories are testimonies of their potential to change the practice and vision of school mathematics teachers. They are powerful statements and reflect the impact of standards-based curricula for teachers at different stages of their careers.

Porath (Chapter 21) and Watts (Chapter 23), both relatively new teachers, described frustration with traditional approaches to teaching. They credit the NSF curriculum materials their schools adopted, coupled with professional development provided through various projects, as the linchpin for major changes and growth in their teaching and student learning. Likewise, Anderson (Chapter 24) underscored the benefits of regular professional development opportunities initiated by state and NSF-funded projects that provided the support and networking needed to sustain and stimulate changes and growth in her teaching. Diliegghio (Chapter 22) described herself as an experienced teacher not satisfied with the outcome of her teaching:

> You have been teaching middle grades mathematics and science for 24 years. The day comes when, although it still feels like you re doing the right thing, it just doesn't seem like you are doing it "right" anymore. You are working hard to prepare and deliver informative lessons. Your students listen... and they try everything you have to offer. How come they often just don't "get it," despite consistent effort on your part and theirs? You think to yourself that surely you can do better and you certainly need something better to offer students.

Her search for a better way led her to a new curriculum that offered some fresh and effective approaches. While she was the lone person in the "reform boat" at her school, the mathematics curriculum materials caused major changes in her teaching approach. She "stopped doing all the work for her students" and began asking herself and her students a new set of questions. The energy level in the classroom rose as did student understanding of mathematics.

All of these teachers described a series of key events in their professional lives including use of NSF curriculum materials coupled with professional development. They saw positive reactions from students as well as questions from parents. These stories represent thousands of others that could be told. They illustrate that curricula can serve as a catalyst for teacher change for new as well as very experienced teachers,

Another personal story of growth stimulated, in part, by the NSF middle school curriculum materials was told by a university mathematician—Ira Papick (Chapter 25). He explained a moment of "light" when he realized that the NSF middle school curriculum materials could help him demonstrate critical links between mathematical ideas within his college course for teachers and concepts embedded in school mathematics. The curriculum materials and his interactions with a group of Missouri middle school teachers provided the impetus for changes in his teaching. It also convinced him of the need for specialized curriculum materials for college-level mathematics courses for preparation of teachers.

Role of University Teacher Educators

Frykholm (Chapter 26) and Bledsoe (Chapter 27) provided practical advice for teacher educators. They related their use of NSF middle school mathematics curriculum materials to engage preservice teachers and to develop their knowledge and skills to teach mathematics. They also emphasized the importance of content knowledge, pedagogical knowledge, belief alignment, as well as the need to comfortably manage standards-based classrooms. Bledsoe emphasized the need to align the methods course with field experiences so that ideas can be reinforced rather than "undone." She also emphasized that before a preservice teacher can change as a teacher, he/she must change as a learner of mathematics. Frykholm illustrated this point by quoting one of his students:

> Had I not taken this class, I probably would go into the classroom and teach the way that I was taught: right out of the book, algorithmically. When children would ask me "Why?" I probably would have said what all of my teachers used to say to me: "Because that's just how you do it. Math has certain rules you just have to learn." How many children are left behind by that approach? But, because of this class, I now feel much more ready to tackle math in my classroom.

The Role of Curriculum Developers

It is noteworthy that several of the chapters in this book were written by authors of the middle school curriculum materials. In traditional curriculum development models, curriculum developers hand over their materials to a publisher and are rarely involved in supporting curriculum implementation. However, the NSF-supported curriculum developers realized early on in their work the importance of initiating teachers and administrators to the materials and providing support to sustain successful implementation.

Since most of the curriculum development projects were not funded to do professional development, the Show-Me Project provided a means not only for providing curriculum-specific professional development but also for bringing together NSF curriculum developers to collaboratively create models and strategies for professional development. In fact, collaboration among curriculum developers working on different sets of materials and for different publishing companies was the hallmark of the Show-Me Project. This type of collaboration is unparalleled in the history of curriculum development in school mathematics.

CLOSING

School administrators, classroom teachers, and parents are looking for better methods of helping students learn important mathematics. While their definition of "important" may differ or be completely undefined in their discussions, they want to improve learning opportunities for students. Through the support of NSF, new curriculum materials exist that provide a core element for school reform. However, most K–12 school systems are unprepared for the increased attention on mathematical reasoning and the pedagogical strategies embedded in standards-based mathematics curricula. Teachers need significant and specialized support. They need to grow mathematically and pedagogically as they experience different ways of learning mathematics.

The Show-Me Project provided support for curriculum adoption and implementation. In some cases this support played out within teachers' classrooms. In other cases, it prompted school districts to launch and sustain long-term reform efforts. The network of professionals (curriculum developers, teacher educators, professional development providers, curriculum researchers) brought together through the Show-Me Project illustrates the power of collaboration and focus. The common thread through all of the work was curriculum materials (textbooks). These items are so core to schooling that they provide the impetus for major change and improvement if coupled with a coordinated effort of professional development, administrator and parent awareness, and documentation of impact. The Show-Me Project provided many of these elements at the critical time when the curriculum materials were being released by the publishers.

History will reveal the long-term impact of the NSF standards-based curriculum development initiative. It is likely that some of the curriculum materials will continue to be used in schools and therefore be visible for many years. Others will influence the next generation of publisher-generated textbooks and therefore will indirectly continue to impact school instruction. The curriculum materials have already influenced thousands of teachers across the country, prompting them to improve their instruction and focus more attention on developing understanding of mathematics. Regardless of the curriculum materials these teachers use in the future, their view of mathematics teaching and student learning will likely be indefinitely altered. The creation and use of mathematics standards-based curriculum materials has prompted unprecedented research activity as well as serious discussion about rigorous methodology. Never before has so much attention, energy, and interest been concentrated on understanding the impact of curriculum materials on student learning and teacher decisions and actions.

While the Show-Me Center will likely fade from memory as funding ceases, the success of NSF's curriculum development initiative is directly related to the work of Centers such as Show-Me that have provided a "place" for scholars, teacher educators, teacher leaders, and mathematicians to come together to discuss and debate issues, to build and modify professional development models, to engage and mentor classroom teachers, and to document and disseminate information.

The Show-Me Center took its name from the Missouri state motto (the Show-Me state). This motto also reflects the spirit of the Center as conveyed by the probable origin of the motto:

> I come from a state that raises corn and cotton and cockleburs... and frothy eloquence neither convinces nor satisfies me. I am from Missouri. You have got to show me. (United States Congressmen from Missouri, Willard Duncan Vandiver in 1899)

To improve opportunities for all students to learn mathematics well, we must move past "frothy eloquence" about what should be done. Teachers and school administrators need tangible tools and methods for improving instruction. NSF's effort to show the field a new way to develop high quality curriculum materials was in that spirit. The Show-Me Center helped support the enactment of the materials and the vision of school mathematics conveyed within the materials.

CHAPTER 29

LOOKING INTO THE FUTURE

Curriculum Challenges for the Next Ten Years

Margaret R. Meyer
University of Wisconsin–Madison, Madison, Wisconsin

Cynthia W. Langrall
Illinois State University, Normal, Illinois

With this chapter we come to the end of this book and the completion of ten years of the Show-Me Project. We have looked back on those ten years with an eye to the lessons learned about curriculum adoption, implementation, professional development, and teacher change. There is the old adage "Hind sight is always 20–20." Indeed it was relatively easy and instructive to look backward. There were many stories to tell and many lessons to make explicit through these stories. Teachers and students have learned mathematics in connected, coherent, and thought-provoking ways from these curricula. The ways in which curricula will be developed in the future will be influenced by the research-based efforts of the authors of these materials. It is not so easy to look forward, beyond the life of the Show-Me Project,

A Decade of Middle-School Mathematics Curriculum Implementation, pages 279–289

to the next ten years for these four NSF-funded middle school curricula and for curriculum reform in general. However, the past ten years and the current state of mathematics education point to several highly probable, inter-related challenges for those interested in improving opportunities for students through curriculum reform.

In this chapter, we present six challenges that teachers, district leaders, teacher educators, or future mathematics curriculum developers might face in the next decade. We do so from our perspective as members of the Show-Me Project for the last ten years. As such, we write about these challenges based on our collective experiences with curriculum-driven reform. Because of the nature of speculation, we ask more questions than we provide answers. It is our hope that by asking questions and making anticipated challenges explicit, we can enable teachers and teacher educators, district leaders, and future curriculum developers to find creative ways to meet the challenges head-on.

CHALLENGE NUMBER 1: MIND THE GAP!

Mathematics for All is a goal that has not been realized despite the attention and efforts of many mathematics educators. What are the effects of the NSF-funded middle grades mathematics curricula on student achievement? Do they help districts narrow the achievement gap among different groups of students? Under what conditions are the curriculum materials effective in providing opportunity for all students to learn significant mathematics? What is the effect of curriculum materials that utilize context for making mathematics accessible for a wide range of students? As an implementation project, Show-Me did not have the resources to collect data to answer these questions explicitly. However, there is growing evidence that in districts where the curriculum materials are used as intended, the entire student population benefits (e.g., Senk and Thompson, 2003; Tarr, Reys, Reys, Chávez, Shih, & Osterlind, 2008; Harwell, Post, Maeda, Davis, Cutler, Anderson & Kahan, 2007; Romberg & Shafer, 2008). But we still need to think about the achievement gap and how the curriculum materials can/should be used in different ways. For example, would it be ok if these curricula improve the achievement of all students but in doing so, widen the gap? These questions point to the broader issue of needing more information about the effects of these curriculum materials on subpopulations. The big questions remain: What are the benefits of these middle school curricula? To what students? Under what implementation conditions?

There is a related gap issue that is revealed when examining the geographical locations of schools that have adopted these curricula. During the course of the Show-Me Project, a visual representation of curriculum adoption sites was created by placing dots, one for each curriculum adoption, on a map of the United States. For the first few years of the project, the satellite directors worked to obtain information from the publisher of their curriculum in order to add dots to the map. This information was surprisingly difficult to obtain for some curricula. There were issues of definition (What amount of money spent on curriculum materials constituted an adoption?), accuracy (Were the sales records complete and up-to-date?), and availability (Given the complex corporate structure and the sensitive nature of the data, who has the authority to release sales data?). Even with these problems, it was exciting to see the map begin to fill in with dots representing adoptions. Not surprisingly, the dots were denser in more populated parts of the country. However, it was discouraging to see large areas of the country with few adoptions. Did these districts not know about these curricula or was it a capacity issue? Were their school boards, administrators, and teachers not ready to consider these materials or were they without sufficient resources to purchase the materials and provide the professional development that teachers required? Whatever the cause, the fact remains that students in many parts of the country have not had the opportunity to learn mathematics using these curriculum materials. *The Show-Me Project is ending, but the job of disseminating quality standards-based mathematics curricula to all students is not done.*

CHALLENGE NUMBER 2: STATE STANDARDS AND ACCOUNTABILITY

The four NSF-funded curricula were developed in response to the 1989 NCTM *Standards.* At that time, few states and fewer districts had a mathematics curriculum framework. The landscape has changed drastically now with most states and districts having their own standards or frameworks. Although it is hard to argue against states and districts thinking about and establishing goals for the mathematical learning of their students, it is unlikely that any one curriculum will match every framework. Will districts refrain from reviewing and adopting curriculum materials that do not align closely with their frameworks? How will districts respond when the carefully designed learning line (trajectory) of a curriculum puts a skill or concept at grade 7 and a state framework demands that it appear in grade 6 or grade 8? Will districts "repackage" curriculum programs in ways that compromise their philosophies, structures, and topic coverage? (For a related discussion, see Challenges 3 and 4.) And finally, a more general question, but one with grave implications: What happens when the expectations for students vary so widely by state and district that students meeting proficiency standards in one state would be considered below proficient in another? The work of the Center for the Study of Mathematics Curricula (Reys, 2006) alerts us to this possibility. Perhaps we should be asking in what way reform curricula might inform state frameworks.

The accountability challenge as embodied in No Child Left Behind (NCLB) is another threat to the long-term viability of the NSF-funded curriculum materials. In some districts teachers and administrative leaders recognize that the goals of NCLB can be achieved by developing students' understanding of mathematics concepts across domains. These districts feel free to adopt and implement these curricula. This is not the case for many low-performing districts where the requirements of annual testing and the potential sanctions for lack of adequate annual progress have turned the focus from learning to teaching-to-the-test. Under these conditions districts attempt to meet the expectations of NCLB by focusing instruction narrowly, and often superficially, only on the concepts and procedures that are embodied in the high-stakes state tests. Adopting one of the NSF-funded curriculum programs is seen as an unjustified gamble and the minimum expectations of basic skills too easily becomes the maximum focus of learning.

Related to these issues is the (mis-)alignment of assessment instruments with district frameworks and the curriculum materials being used in the classroom. There are three possible mismatches among these factors, each of which will doom even the best curriculum implementation. First of all, if the district framework is not aligned with the assessment, there is no way to know if the district goals are being met. Secondly, if the district framework is not aligned with the curriculum materials provided by the district, then teachers must create their own supplemental materials. An exception to this occurs when the curriculum goals exceed those of the district framework, as is often the case when the district is using one of the four NSF-funded curricula. Finally when the assessment is not aligned with the curriculum, students will either be tested on mathematical content they have not had the opportunity to learn or they will be tested on only a small part of what they have learned. This latter situation occurs when the assessment focuses on basic skills and the curriculum addresses not only basic skills but also problem solving, communication, and connections. In the past the Show-Me Project has worked to combat the tensions that arise with regard to curriculum alignment and accountability issues. *The Show-Me Project is ending, but the challenge of implementing standards and accountability to improve student learning remains.*

CHALLENGE NUMBER 3: THE PUBLISHERS' PERSPECTIVE AND THE BOTTOM LINE

Working with publishers has been the ultimate "good news–bad news" story since the early days of the development of these NSF-funded curriculum materials. First, publishers put their stamp on the curriculum materials. For each curriculum, they contributed a polished layout, attractive artwork, skillful editing, and a professional production team. Secondly, each publisher provided a national marketing plan and sales force. The curricula would never have achieved national exposure without the work of the publishers. NSF knew this was important when they made publisher contracts a condition of funding for each curriculum development project. Third, publishers have been willing to learn what it takes to support standards-based curricula. This has not been an easy lesson because the professional development model is so different from that which publishers have historically done to familiarize teachers with their mathematics curricula. Finally, the publishers have been supportive of the work of the Show-Me Project. They have provided sample materials for the curriculum showcases and follow-up to questions about their particular curriculum. The collaboration between Show-Me and the publishers has been a "win-win" situation. Show-Me provided the publishers with valuable district contacts and eventual sales through the showcases and the publishers in turn provided materials for the showcases and professional development following adoptions.

There have also been many frustrations connected with working with the publishers. Each of the curriculum developers has stories to tell about the publishing process. Here are examples that illustrate some of those frustrations that curriculum developers have faced:

- In one instance, publishing company editors did not understand the intent of mathematical problems as written by the curriculum developers and changed them in ways they thought would be less ambiguous to the students. As a result, high-level problem-solving situations became low-level recall problems. The curriculum developers had to be very vigilant in reviewing their materials throughout the publication process to catch changes of this type.
- A publisher tried to respond to demands for more practice by developing skill and drill sheets that were antithetical to the philosophy and pedagogical approach of the NSF-funded curricula.
- At a publisher's exhibit at a national conference, only the mathematics curricula that had been developed in-house, and by the publisher, were displayed. Teachers had to ask to see the NSF-funded curriculum materials, which were being kept in boxes under a table.
- A publisher had two competing middle school mathematics curricula. One was a traditional middle school mathematics textbook series, which required little publisher-sponsored professional development. The other was one of the NSF-funded curricula, which required extensive publisher-sponsored professional development.

Thus the promotion of one of the curricula was more cost-effective for the publishing company.

- The original publishing company was bought by another publishing company, resulting in the curriculum developers having to establish new relationships with the employees at the new company. This situation added months to the curriculum development timeline.
- A publisher sales representative was not knowledgeable about the strengths of the NSF-funded curriculum. The publishing company's traditional textbook series was easily marketable, so the sales representative promoted the traditional mathematics textbook series and offered to give away copies of the NSF-funded curriculum materials as a premium. Such giveaways made it difficult to track sales, thus impacting the market share of NSF-funded curriculum materials.

One thing is clear: in the real world of business one must keep an eye on the bottom line. It is a fact of life that if the curriculum does not sell, the publisher will not continue to support it. What are some of the pressures that publishers feel in this current climate of state standards and accountability?

- Schools often require a new copyright on materials that they will consider for adoption. In addition, there is a continual race with the competing curricula to add on many curricular "go-withs," such as test generators, on-line support, manipulative kits, and so on. All of these cost money and sometimes they are not consistent with the philosophy of the curriculum. However, if the publisher does not make these available, the curriculum might not be adopted, because the competition does have them.
- There is a move toward customization of curricula in order to meet state standards or even customization to meet particular student needs in large districts. Publishers have started to accommodate these expectations with on-demand publishing formats. This move could have a damaging effect on the coherence of these four curricula as the publishers/districts pick and choose units or parts of units to match states' textbook adoption list, testing cycles, and perceived student needs. The ebb and flow of goals established in local and state standards documents clearly favor traditional textbook materials that are readily adaptable, but as a consequence superficial as compared to the NSF-funded materials that reflect careful development and field-testing.
- Publishers often try to bundle their own commercially developed program with a NSF-funded curriculum. They see this as a way to sell both of the curricula they own and to provide skills building, concept

building, and problem solving in response to state standards. The effect is that teachers have to manage two very different pedagogical styles. The students are left wondering which program is the "real mathematics." Teachers who are reluctant to change their practice are able to continue to focus on the traditional.

- Textbook adoption states drive the publishing world with the promise of potentially huge markets. However, the cost to publishers of entering the competition for state approval is enormous. Some publishers are reluctant to make that investment, especially when the state frameworks are often so restrictive that standards-based curricula have little chance of being approved.

The world of textbook publishing is small and getting smaller. Through the Show-Me sponsored curriculum showcases, many of the publisher sales representatives became familiar with all of the NSF-funded curriculum materials. This was a good thing because it allowed them to understand common themes across the curricula. It also meant that in the future when they might be working for the competition, they were ready to go. This was an indirect benefit of Show-Me, a value added that will not continue now that the project is ending.

In concluding this section on the publisher–developer relationship, it should be noted that the Show-Me Project provided an extended contact link between the publishers and the developers at the satellites. This contact provided expertise to the publisher about philosophy, intent, and professional development. As a result, many commercially developed books have been revised to look more like the standards-based texts. *The Show-Me Project is ending, but the need for continued contact and cooperation between the publishers and the developers remains.*

CHALLENGE NUMBER 4: MAINTAINING CURRICULUM BALANCE AND INTEGRITY

Over the course of the ten years of the Show-Me Project, the four sets of curriculum materials evolved in response to several outside forces. The most obvious outside force influencing the curricula has been the state standards movement. One effect of state standards has been to push the first-year algebra course down into the middle school. At issue is agreement about what constitutes a first-year algebra course. All four of the NSF-funded middle school mathematics curriculum materials contained conceptually rich algebra strands in their first editions and these strands were further developed in the revisions. They develop the conceptual foundations of algebra, including variation, variables, patterning, and graphing.

Although algebra content increased in the revised materials, all of the curricula advocate a balance of curricular topics at the middle grades as proposed by the 1989 *Standards* and affirmed by the 2000 *Principles and Standards for School Mathematics.*

The recent publication of the NCTM *Focal Points* (2006) provides another pressure. The curricula were written with the NCTM curriculum standards grade bands in mind (1989, 2000). These grade bands provided particular content to be achieved by the end of the last grade in the band. The *Focal Points* identify content as the curricular focus of specific grades. If these recommendations are followed, there will be problems with the NSF-funded curricula, which identify well articulated learning trajectories of key content to be developed over time, where the formalization of concepts is not always the first goal at a particular grade level in which the content is taught. It will be too easy for the *Focal Points* document to be misinterpreted and the minimum expectations at each grade level could become the maximum. The result will be a demand for curricula that are much narrower and less coherent than the current standards-based curricula.

Another source of pressure comes from the vocal opponents of the NSF-funded curricula. The so-called "math wars" have raged since the beginning of the Show-Me Project and there seems to be little hope that they will end while the curricula are still being adopted and used. NSF-funded curricula at all grade levels have been targeted. At the middle grades, curricula have been attacked particularly on the basis of long standing expectations for students to have acquired certain skills, mostly involving number, by the end of grade eight. It is amazing to note the number of news articles fanning the flames of disagreement over the best approach to teach middle school mathematics. These four curricula (and others) have been branded as being "constructivist," "self-discovery," and containing "fuzzy math." Their critics have made these phrases into a rallying cry for the development of skill in computing with "standard" algorithms as the primary goal of mathematics classes, ignoring conceptual development, problem solving skills, and the development of mathematical thinking habits of mind. For the past ten years the Show-Me Project has provided leadership and clarity of message in response to these attacks in a way that the individual curriculum developers alone could not. The Project provided a core group of knowledgeable leaders who could argue for the need for balance among curricular topics and respond to criticisms in articulate and impartial ways. *The Show-Me Project is ending, but the need for an impartial analysis of what these curricula embody and advocacy for developmentally-appropriate instructional materials continues.*

CHALLENGE NUMBER 5: TEACHER CAPACITY

Teacher capacity for using the NSF-funded curriculum materials effectively remains the biggest challenge for the next decade. There are at least two dimensions to this challenge: teacher (and district) capacity to sustain professional development and teacher fidelity to the curriculum materials.

Since the publication of the 1989 *Standards* it has been clear that teachers are being asked to teach different mathematics (than they learned) in different ways (than they were taught) and this demands significant professional development. The advent of the standards-based curricula highlighted the need for teachers to become life-long learners of mathematics at the same time as they are teachers. Chapters in this book show how curricula can have a profound impact on preservice and in-service teachers' growth and development. These stories point to the power of communities of practice as a way to sustain a commitment to continual professional development. After a decade of the Show-Me Project, the first students of these curricula could be emerging as new middle school teachers. We hope their experiences as learners of mathematics with these curricula were strong enough that they remember the joy of understanding mathematics when they begin to teach from these materials in their own classrooms.

The need for sustained professional development is crucial in supporting teachers as life-long learners. However, in an effort to jump start lagging professional development, we have seen many cases where districts periodically switch from one NSF-funded curriculum to another. Although there may be good reasons for doing this, such as promoting a boost in teacher excitement gained from starting something new, the end result could be damaging if repeated adoption cycles through NSF-funded programs leads teachers to perceive each change to a new curriculum as a failure of the previous one.

The second dimension of teacher capacity is teacher fidelity to the curricula. How do we know the curricula work if teachers are not teaching the materials in the way the developers intended? There are at least three reasons why teachers might not teach a curriculum as its developers intended. One reason has to do with not knowing the developers' intentions. Another reason might be that teachers are not comfortable teaching in this way, so they stay within their comfort level and teach the materials as they have taught previously. A third reason for lack of fidelity is a pervasive culture in education that tells teachers that they should be teaching from a district defined framework using district developed packaged units rather than using a textbook. The director of instructional services for a very large urban district was recently heard to say "We have to get the teachers away from those textbooks." The implication was clear that good instruction could not rely on textbook materials. At one time, before the availability of NSF-funded curricula, adapting or abandoning textbook materials was necessary. How do we now encourage curriculum fidelity and at the same time support teachers as professionals who make informed decisions about instruction in their classrooms? How do we highlight the growth in mathematical understanding and pedagogy that is possible when teachers use instructional materials that are "teacher serious" rather than "teacher friendly?" For the past decade the Show-Me Project has been demonstrating the value of professional development to support the fidelity of curriculum implementation and the importance of curriculum materials to support teachers in their classrooms. *The Show-Me Project is ending, but there are many who still need to learn how to support curriculum implementation through professional development that helps teacher grow.*

CHALLENGE NUMBER 6: WHO WILL TAKE UP THE WORK THAT REMAINS?

With its funding from NSF, the Show-Me Project provided a collective legitimacy to the four sets of curriculum materials. This was helpful to districts hoping to utilize curriculum materials to improve instruction. Districts were able to go to Show-Me curriculum showcases or to the Show-Me website and do a first pass evaluation of the four curricula. In this way the Show-Me Project provided both a central clearinghouse for the most current information about the materials and contact information for direct links to the developers, a subset of whom received support from the Project so they were available to respond directly to inquiries about the curricula. This direct link to the developers was rather extraordinary in an age where there are usually several layers between consumers and developers of any product. Without the Show-Me Project, the link will now be between the districts and the publishers. The promotion of curricula will be dependent on the sales representatives in various localities and their general awareness of the curricula, which is often minimal. The challenge for district administrators will be to recreate the showcase context in order to decide what materials will be reviewed for possible adoption.

With the end of the Show-Me Project who will take up the work of curriculum development and support in the next decade? Many of the people currently involved in curriculum development may not be actively involved in curriculum development or supporting implementation ten years from now. The task of curriculum development and support will belong to younger people as yet unidentified. That fact raises other questions. Under what conditions will NSF fund mathematics curriculum development in the future? Will the next generation of mathematics curricula be produced by the remaining few publishers, who are simply responding to new policies and market demands? Will any of these four middle grades curricula still be commercially viable in ten years or will they be historical artifacts studied by mathematics education doctoral students trying to understand the period of education reform launched by the 1989 *Standards*? Will there be funds from NSF or other sources to support the professional development needs of our teachers or will this become the sole responsibility of the districts, which are often dependent on the unreliable economy and tight budgets of their states?

And finally, what will developers of future curricula and providers of professional development take away from *a decade of middle school mathematics curriculum implementation and the lessons learned from the Show-Me Project*? Time will tell. Maybe ten years from now copies of this book will provide affirmation, inspiration, and challenge: affirmation that the Show-Me Project support of the NSF-funded middle grades mathematics curricula made a significant difference in teacher and student learning; inspiration from the continued stories of districts, teachers, and students who will experience these materials; and challenge to find ways to continue the work of the Show-Me Project.

REFERENCES

Acquarelli, K., & Mumme, J. (1996). A renaissance in mathematics education reform. *Phi Delta Kappan, 77,* 478.

American Association for the Advancement of Science. (2000). *Middle grades mathematics textbooks: A benchmarks-based evaluation.* Washington, DC: American Association for the Advancement of Science.

Ball, D. L., & Cohen, D. K. (1996). Reform by the book: What is—or might be—the role of curriculum materials in teacher learning and instructional reform? *Educational Researcher, 25,* 6–8, 14.

Ball, D.L., Ferrini-Mundy, J., Kilpatrick, J., Milgram, R.J., Schmid, W., & Schaar, R. (2005). Reaching for common ground in K–12 mathematics education. *Notices of the American Mathematical Society, 52*(9), 1055–1058.

Bay, J. M. (1999). *Middle-school mathematics curriculum implementation: The dynamics of change as teachers introduce and use standards-based curricula.* Unpublished doctoral dissertation, University of Missouri-Columbia.

Bay, J., Reys, B., & Reys, R. (1999). Implementation of standards-based mathematics curriculum: Top 10 elements that must be in place. *Phi Delta Kappan, 80,* 503–506.

Becker, J. P., & Jacob, B. (2000a). The politics of California school mathematics: The anti-reform of 1997-99. *Phi Delta Kappan, 81,* 529–537.

Becker, J. P., & Jacob, B. (2000b). Look at the details: A reply to Deborah Haimo and James Milgram. *Phi Delta Kappan, 82,* 147–148.

Beem, J. (2005). *Geometry connections.* Upper Saddle River, NJ: Prentice Hall.

Billstein, R., & Williamson, J. (1999a). *Middle grades MATHThematics (Books 1–3).* Evanston, IL: McDougal Littell.

Billstein, R., & Williamson, J. (1999b). *Middle grades MATHThematics teacher resource books.* Evanston, IL: McDougal Littell.

Bolster Education. (2001). *MMM: Modeling middle school mathematics* (video series). New York: Thirteen/WNET New York.

A Decade of Middle-School Mathematics Curriculum Implementation, pages 291–298

Bolster, C. & Reys, R. E. (2002). Modeling middle school mathematics: A technology-based professional development resource. *Journal of Mathematics Education Leadership, 6*(1), 36–39.

Borko, H., & Putnam, R. T. (1996). Learning to teach. In D. C. Berliner and R. C. Calfee (Eds.), *Handbook of educational psychology* (pp. 673–708). New York: McMillan.

Bousquet J., Janssen S., & Brodesky, A. (2005). Buyer beware: Rates, ratios, percents, and proportions. In *MathScape: Seeing and thinking mathematically, course 2* (pp. 6–45). New York: Glencoe/McGraw Hill.

Brodesky A., Brown R., Tobin D. (2005). What does the data say?: Graphs and averages. In *MathScape: Seeing and thinking mathematically, course 1 teacher's guide* (p. A19). New York: Glencoe/McGraw Hill

Brown, J., Collins, A., & Duguid, P. (1989). Situated cognition and the culture of learning. *Education Researcher, 18*(1), 32–42.

Brown, S. I. (1993). Towards a pedagogy of confusion. In A. M. White (Ed.*), Essays in humanistic mathematics* (pp. 107–122). Washington, DC: Mathematical Association of America.

Brown, S., Cooney, T., & Jones, D. (1990). Mathematics teacher education. In W. R. Houston (Ed.), *Handbook of research on teacher education* (pp. 639–656). New York: Macmillan.

Brutlag, D. (2005). Roads and ramps: Slopes, angles and ratios. In *MathScape: Seeing and thinking mathematically, course 3* (p. 262). New York: Glencoe/McGraw Hill.

Burrill, G. (1998). Changes in your classrooms: From the past to the present to the future. *Mathematics Teaching in the Middle School, 4*, 184–190.

Carroll, W. M. (1997). Results of third-grade students in a reform curriculum on the Illinois state mathematics test. *Journal for Research in Mathematics Education, 28*, 237–242.

Cole, B., Romberg, T. A., de Lange, J., & Foster, S. (1998). Comparing quantities. In National Center for Research in Mathematical Sciences Education & Freudenthal Institute (Eds.) *Mathematics in context.* Chicago, IL: Encyclopaedia Britannica.

Cole, B. Romberg, T. A., de Lange, J., & Foster, S. (1998). Teacher resource and implementation guide. In National Center for Research in Mathematical Sciences Education & Freudenthal Institute (Eds.) *Mathematics in context.* Chicago, IL: Encyclopaedia Britannica.

COMAP. (1998). *Mathematics: Modeling our world. Courses 1–3.* Cincinnati: South-Western Publishing Company.

Conference Board of Mathematical Science. (2001). *The mathematical education of teachers report.* Washington, DC: Mathematical Association of America.

Cooney, T. J. (1994). Teacher education as an exercise in adaptation. In D. B. Aichele & A. F. Coxford (Eds.), *Professional development for teachers of mathematics: 1994 yearbook* (pp. 9–22). Reston, VA: NCTM.

Coxford, A. F., Fey, J. T., Hirsch, C. R., Schoen, H. L., Burrill, G., Hart, E. W., & Watkins, A. E. (1998). *Contemporary mathematics in context: A unified approach (courses 1–4).* New York: Glencoe/McGraw-Hill.

Cuoco, A., Goldenberg, E. P., & Mark, J. (1996). Habits of mind: An organizing principle for mathematics curricula. *Journal of Mathematical Behavior, 15*, 375–402.

Dekker, T. (1997). *Enquiry project: Assessing realistic mathematics education.* Unpublished manuscript. Utrecht, The Netherlands: Freudenthal Institute.

Denny, R., & Williamson, J. (1999). *Middle grades MATHThematics professional development handbook (books 1–3).* Evanston, IL: McDougal Littell.

Donovan, E., Bransford, J., & Pellegrino, J. (2000). *How people learn.* Washington, DC: National Academy Press.

Dow, P. (1991). *Schoolhouse politics: Lessons from the Sputnik era.* Cambridge, MA: Harvard University Press.

Echevarria, J., Vogt, M., & Short, D. J. (2004). *Making content comprehensible for English language learners: The SIOP model.* Boston: Allyn and Bacon.

Education Development Center, Inc. (1998). *MathScape: Seeing and thinking mathematically.* Mountain View, CA: Creative Publications.

Education Development Center, Inc. (2005). *MathScape: Seeing and thinking mathematically.* New York, NY: Glencoe/McGraw Hill.

Ellington, A. J. (2003). A meta-analysis of the effects of calculators on students' achievement and attitude levels in precollege mathematics classes. *Journal for Research in Mathematics Education, 34*, 433–463.

Fendel, D., Resek, D., Alper, L., & Fraser, S. (1997–1999). *Interactive mathematics program: Integrated high school mathematics, year 1–4.* Berkeley, CA: Key Curriculum Press.

Finn, E., & Ravitch, D. (2004). *The mad, mad world of textbook adoption.* Washington, DC: Thomas B. Fordham Institute.

Forman, E. A. & McPhail, J. (1993). Vygotskian perspective on children's collaborative problem-solving activities. In E. A. Forman, N. Minick, & C. A. Stone (Eds). *Interpersonal relations in formal and informal education* (pp. 213–229). New York: Oxford University Press.

Frykholm, J. A. (1996). Preservice teachers in mathematics: Struggling with the standards. *Teaching and Teacher Education, 12*, 665–681.

Frykholm, J. A. (1998). Beyond supervision: Learning to teach mathematics in community. *Teaching and Teacher Education, 14*, 305–322.

Frykholm, J. A. (2000). Elementary mathematics: The missing piece in secondary teacher preparation? *FOCUS on Learning Problems in Mathematics, 22*(1), 27–44.

Garrison, L., & Mora, J. (1999). Adapting mathematics instruction for English-language learners: The language–concept connection. In L. Ortiz-Franco, N. G. Hernandez, & Y. De La Cruz (Eds.), *Changing the faces of mathematics: Perspectives on Latinos* (pp. 35–47). Reston, VA: National Council of Teachers of Mathematics.

Goldsmith, L. T., Mark, J., & Kantrov, I. (1998). *Choosing a standards-based mathematics curriculum.* Boston, MA: Educational Development Center, Inc.

Harcharras, A., & Mitrea, D. (2006). *Calculus connections.* Upper Saddle River, NJ: Prentice Hall.

Harwell, M. R., Post, T. R., Maeda, Y., Davis, J. D., Cutler, A. L., Anderson, E., & Kahan, J. A. (2007). Standards-based mathematics curricula and secondary students' performance on standardized achievement tests. *Journal for Research in Mathematics Education, 38*, 71–101.

Heaton, R. M. (2000). *Teaching mathematics to the new standards: Relearning the dance.* New York: Teachers College Press.

Henningsen, M., & Stein, M. K. (2002). Supporting students' high-level thinking, reasoning, and communication in mathematics. In J. Sowder & B. Schappelle (Eds.) *Lessons learned from research* (pp. 27–36). Reston, VA: National Council of Teachers of Mathematics.

Her, T., & Webb, D. C. (2004). Retracing a path to assessing for understanding. In T. A. Romberg (Ed.), *Insight stories: Assessing middle school mathematics* (pp. 200–220). New York: Teachers College Press.

Herman, J. L., & Abedi, J. (2004). *Issues in assessing English language learners' opportunities to learn mathematics.* Center for the Study of Evaluation (CSE) Report 633. Los Angeles, CA: National Center for Research on Evaluation, Standards, and Student Testing (CRESST).

Hiebert, J., Carpenter, T. P., Fennema, E., Fuson, K. C., Wearne, D., Murray, H., Human, P., & Olivier, A. (1997). *Making sense: Teaching and learning mathematics with understanding.* Portsmouth, NH: Heinemann.

Hill, H., Rowan, B., & Ball, D. L. (2005). Effects of teachers' mathematical content knowledge for teaching on student achievement. *American Educational Research Journal, 42,* 371–76.

Hirsch, C. R., (Ed.). (2007). *Perspectives on the design and development of school mathematics curricula.* Reston, VA: National Council of Teachers of Mathematics.

Hord, S. M., Rutherford, W. L., Huling-Austin, L. & Hall, G. E. (1998). *Taking charge of change.* Austin, TX: Southwest Educational Development Library.

Huntley, M. A., Rasmussen,C. L., Villarubi, R. S., Sangtong, J., & Fey, J. T. (2000). Effects of standards-based mathematics education: A study of the *Core-Plus* mathematics project algebra and functions strand. *Journal for Research in Mathematics Education, 31,* 328–361.

Karnowski, E., Benson, S., Janssen, S., Doherty, A., Fagan, E., & Thorpe Breeze, C. (2005). Beside the point: Operating with decimals, percents, and integers. In *MathScape: Seeing and thinking mathematically, course 1* (pp. 210–275). New York: Glencoe/McGraw Hill.

Karnowski, E., Fagan, E., Thorpe Breeze, C., & Marrero, J. (2005). From wholes to parts: Operating with factors, multiples, and fractions. In *MathScape: Seeing and thinking mathematically, course 1* (pp. 96–161). New York: Glencoe/McGraw Hill.

Kenney, J. M. (2005). Mathematics as language. In J. M. Kenney, E. Hancewicz, L. Heuer, D. Metsisto, & C. L. Tuttle (Eds.) *Literacy strategies for improving mathematics instruction* (pp. 1–8). Alexandria, VA: Association for Supervision and Curriculum Development.

Khisty, L. L. (1997). Making mathematics accessible to Latino students: Rethinking instructional practice. In J. Trentacosta & M. Kenny (Eds.), *Multicultural and gender equity in the mathematics classroom: The gift of diversity, 97th yearbook of the National Council of Teachers of Mathematics* (pp. 92–101). Reston, VA: National Council of Teachers of Mathematics.

Kilpatrick, J. (1997). Confronting reform. *American Mathematical Monthly, 104,* 955–962.

Kindt, M., Abels, M., Dekker, T., Meyer, M. R., Pligge M. A., & Burrill, G. (2006). Comparing quantities. In Wisconsin Center for Education Research & Freudenthal Institute (Eds.), *Mathematics in context.* Chicago: Encyclopaedia Britannica.

Kindt, M., Abels, M., Meyer, M. R., & Pligge, M. A. (1998). Comparing quantities (teacher edition). In National Center for Research in Mathematical Sciences Education & Freudenthal Institute (Eds.) *Mathematics in context.* Chicago, IL: Encyclopaedia Britannica, 98.

Kindt, M., Abels, M., Meyer, M. R., & Pligge, M. A. (1998). Comparing quantities. In National Center for Research in Mathematical Sciences Education & Freudenthal Institute (Eds.) *Mathematics in context.* Chicago, IL: Encyclopaedia Britannica, 25.

Kindt, M., Abels, M., Meyer, M.R., & Pligge, M.A. (1998). Teacher resource and implementation guide. In National Center for Research in Mathematical Sciences Education & Freudenthal Institute (Eds.) *Mathematics in context.* Chicago, IL: Encyclopaedia Britannica.

Kindt, M., Dekker, T., Burrill, G., & Romberg T. A. (2006). Number tools. In Wisconsin Center for Education Research & Freudenthal Institute (Eds.), *Mathematics in context.* Chicago: Encyclopaedia Britannica, Inc.

Kleiman G., Fagan E., Janssen S., Brodesky A., & Tobin D. (2007). The case of *MathScape*: Seeing and thinking mathematically in the middle grades. In C. Hirsch (Ed.). *Perspectives on design and development of school mathematics curricula* (pp. 95–106). Reston, VA: National Council of Teachers of Mathematics.

Kohn, A. (2001). Fighting the tests. *Phi Delta Kappan, 82,* 349–357.

Lappan, G., Fey, J. T., Fitzgerald, W., Friel, S. N., & Phillips, E. D. (1998, 2002, 2004). *Connected mathematics series.* Needham, NJ: Prentice Hall.

Lappan, G., Fey, J. T., Fitzgerald,W., Friel, S. N., & Phillips, E. D. (2006). *Connected mathematics 2.* Boston, MA: Pearson-Prentice Hall.

Lindsey, V. (1926). *The leaden-eyed, collected poems.* New York: D. Appleton and Company.

Lloyd, G. M., & Frykholm, J. A. (2000). How innovative middle school mathematics materials can change prospective elementary teachers' conceptions. *Education, 120,* 575–580.

Loucks-Horsley, S., Love, N., Styles, K. E., Mundry, S., & Hewson, P. W. (2003). *Designing professional development for teachers of science and mathematics* (2nd ed.). Thousand Oaks, CA: Corwin Press.

Ma, L. (1999). *Knowing and teaching elementary mathematics: Teachers' understanding of fundamental mathematics in China and the United States.* Mahwah, NJ: Lawrence Erlbaum.

Manouchehri, A. (1997). School mathematics reform: Implications for mathematics teacher preparation. *Journal of Teacher Education, 48,* 197–209.

Mathematical Sciences Education Board and National Research Council (1989). *Everybody counts: A report to the nation on the future of mathematics education.* Washington, DC: National Academy Press.

McIntosh, M. E. (1997). Formative assessment in mathematics. *Clearing House, 71,* 92–97.

Meyer, M. R. (2006). Teacher implementation guide. In Wisconsin Center for Education Research & Freudenthal Institute (Eds.) *Mathematics in context*, Chicago, IL: Encyclopaedia Britannica.

Meyer, M. R. (1999). Multiple strategies = multiple challenges. *Mathematics Teaching in the Middle School, 4*, 519–523.

Meyer, M. R., Dekker, T. & Querelle, N. (2001) Context in mathematics curricula. *Mathematics Teaching in the Middle School, 6*, 522–527.

Moschkovich, J. N. (1999). Understanding the needs of Latino students in reform-oriented mathematics classrooms. In L. Ortiz-Franco, N. G. Hernandez, & Y. De la Cruz (Eds.), *Changing the faces of mathematics: Perspectives on Latinos* (pp. 5–12). Reston, VA: National Council of Teachers of Mathematics.

National Center for Research in Mathematical Sciences Education & Freudenthal Institute. (Eds.). (1997, 1998, 2006). *Mathematics in context.* Chicago: Encyclopaedia Britannica.

National Commission on Excellence in Education. (1983). *A nation at risk: The imperative for educational reform.* Washington, DC: U. S. Government Printing Office.

National Council of Teachers of Mathematics. (1989). *Curriculum and evaluation standards for school mathematics.* Reston, VA: Author.

National Council of Teachers of Mathematics. (1991). *Professional standards for teaching mathematics.* Reston, VA: Author.

National Council of Teachers of Mathematics. (1995). *Standards for mathematics assessment.* Reston, VA: Author.

National Council of Teachers of Mathematics. (2000). *Principles and standards for school mathematics.* Reston, VA: Author.

National Council of Teachers of Mathematics. (2006). *Curriculum focal points for prekindergarten through grade 8 mathematics: A quest of coherence.* Reston, VA: Author.

National Research Council. (1989). *Everybody counts: A report to the nation on the future of mathematics education.* Washington, DC: National Academy Press.

National Science Foundation (2006). *National science foundation investing in America's future: Strategic plan FY 2006–2011.* Retrieved http://www.nsf.gov, August 8, 2008.

Panasuk, R. M., Cutler, J., & Sullivan, M. M. (1998). Enhancing the field experience component in a mathematics teacher preparation program. *NCSM Journal of Mathematics Education Leadership, 1*(4), 16–23.

Papick, I. (2006). *Algebra connections.* Upper Saddle River, NJ: Prentice Hall.

Perkins, I., & Flores, A. (2002). Mathematical notations and procedures of recent immigrant students. *Mathematics Teaching in the Middle School, 7*, 346–351.

Perkowski, D., & Perkowski, M. (2006). *Data analysis and probability connections.* Upper Saddle River, NJ: Prentice Hall.

Piaget, J. (1952). *The origins of intelligence in children.* New York: International University Press.

Polya, G. (1973). *How to solve it: A new aspect of mathematical method.* Princeton, NJ: Princeton University Press.

Pugalee, D. K. (2005). *Writing for mathematical understanding.* Norwood, MA: Christopher Gordon Publishers.

Reed, C. F. (2004). Mathematically gifted in the heterogeneously grouped mathematics classroom: What is a teacher to do? *Journal of Secondary Gifted Education, 15*(3), 89–95.

Reys, B. J. (Ed.) (2006). *The intended mathematics curriculum as represented in state-level curriculum standards: Consensus or confusion?* Greenwich, CT: Information Age.

Reys, B. & Reys, R. (2006). The development and publication of elementary mathematics textbooks: Let the buyer beware. *Phi Delta Kappan, 87,* 377–383.

Reys, B. J., Reys, R. E., & Chavez, O. (2004). Why mathematics textbooks matter. *Educational Leadership, 61*(5), 61–66.

Reys, B. J., Reys, R. E., Barnes, D., Beem, J. K., & Papick, I. (1997). Collaborative curriculum review as a vehicle for teacher enhancement and mathematics curriculum reform. *School Science and Mathematics, 97,* 253–259.

Reys, B. J., Robinson, E., Sconiers, S. & Mark, J. (1999). Mathematics curricula based on rigorous national standards: What, why and how? *Phi Delta Kappan, 80,* 454–456.

Reys, R. E. (2001a). Mathematics curricula and guinea pigs. *Mathematics Teacher, 94,* 6.

Reys, R. E. (2001b). Curricular controversy in the math wars: A battle without winners. *Phi Delta Kappan, 83,* 255–258.

Romberg, T.R., & Shafer, M. (Eds.) (in press). *The impact of reform mathematics instruction on student achievement.* Mahwah, NJ: Lawrence Erlbaum.

Russell, S., Tierney, C., Mokros, J. and Goodrow, A. (1997). *Investigations in number, data, and space.* Palo Alto, CA: Dale Seymour.

Schoenfeld, A. H. (2004). The math wars. *Educational Policy, 18*(1), 253–286.

Senk, S. L., & Thompson, D. R. (2003). *Standards-based school mathematics curricula: What are they? What do students learn?* Mahwah, NJ: Lawrence Erlbaum.

Sheffield, L. J. (Ed.). (1999). *Developing mathematically promising students.* Reston, VA: National Council of Teachers of Mathematics.

Shulman, L. (1986). Paradigms and research programs in the study of teaching: A contemporary perspective. In M. Wittrock (Ed.) *Handbook of research on teaching* (Vol. 3, pp. 3–36). New York: Macmillan.

Sjoberg, C. A., Slavit, D., & Cooney, T. (2004). Improving writing prompts to improve student reflection. *Mathematics Teaching in the Middle School, 9,* 490–493.

Stigler, J., & Hiebert, J. (1999). *The teaching gap: Best ideas from the world's teachers for improving education in the classroom.* New York: Simon and Schuster.

Tarr, J. Reys, R., Reys, B., Chávez, O., Shih, J., & Osterlind, S. (2008) The impact of middle grades mathematics curricula on student achievement and the classroom learning environment. *Journal for Research in Mathematics Education, 39,* 247–280.

Tepper-Haimo, D., & Milgram, J. (2000). Professional mathematicians comment on school mathematics in California, *Phi Delta Kappan, 82,* 145–46.

Tetley, L. (1998). Implementing change: Rewards and challenges. *Mathematics Teaching in the Middle School, 4,* 160–164.

Tyson-Bernstein, H. (1988). *A conspiracy of good intensions: America's textbook fiasco,* Washington, DC: Council for Basic Education.

Vacc, N. N. & Bright, G. W. (1994). Changing preservice teacher-education programs. In D. B. Aichele & A. F. Coxford (Eds.), *Professional development for teach-*

ers of mathematics: 1994 yearbook (pp. 116–127). Reston, VA: National Council of Teachers of Mathematics.

Van Boening, L. (1999). Growth through change. *Mathematics Teaching in the Middle School, 5,* 27–33.

Van Galen, F., Abels, M., Burrill, G., Spence, M.S., & Hedges, T. (2006). Some of the parts. In Wisconsin Center for Education Research & Freudenthal Institute (Eds.), *Mathematics in context.* Chicago: Encyclopaedia Britannica, Inc.

Van de Walle, J. A. (2000). *Elementary and middle school mathematics: Teaching developmentally, 4th edition.* New York: Longman.

Villaume, S.K. (2000). The necessity of uncertainty: A case study of language arts reform. *Journal of Teacher Education, 51*(1), 18–25.

Wagreich, P., Goldberg, H., Beiler, J., & Simpson Beissinger, J. (1997). *Math trailblazers.* Dubuque, IA: Kendall/Hunt Publishing Company.

Webb, D. C., Krusi, J., Palha, S., & Dekker, T. (2006). Patterns and figures [Teacher Guide]. In Wisconsin Center for Education Research & Freudenthal Institute (Eds.), *Mathematics in context.* Chicago: Encyclopaedia Britannica, Inc.

Webb, D. C., & Meyer, M. R. (2007). The case of *Mathematics in Context.* In C. Hirsch (Ed.), *Perspectives on design and development of school mathematics curricula* (pp. 81–94). Reston, VA: National Council of Teachers of Mathematics.

Williams, J. A. (2001). Classroom conversations; Opportunities to learn for ESL students in mainstream classrooms. *Reading Teacher, 54,* 750–757.

Wisconsin Center for Education Research & Freudenthal Institute (Eds.) (2006). *Mathematics in context.* Chicago: Encyclopaedia Britannica.

Woodward, J., & Howard, L. (1994). The misconceptions of youth. Errors and their mathematical meaning. *Exceptional Children, 61,* 126–136.

Wooten, W. (1965). *SMSG: The making of a curriculum,* New Haven: Yale University Press.

Wu, H. H. (1997). The mathematics education reform: Why you should be concerned and what you can do. *American Mathematical Monthly, 104,* 946–54.

Zawojewski, J. S., Robinson, M., & Hoover, M. (1999). Innovations in curriculum: Reflections on developing formal mathematics and the *Connected Mathematics Project. Mathematics Teaching in the Middle School, 4,* 324–331.

Zeichner, K. (1993). *Designing educative practicum experiences for prospective teachers.* Paper presented at the International Conference on Teacher Education: From Practice to Theory, Tel-Aviv, Israel.

www.ingramcontent.com/pod-product-compliance
Lightning Source LLC
Chambersburg PA
CBHW051150180926
51611CB00001B/13
9781607520122